D0409074

CENTAURI DREAMS

CENTAURI DREAMS

IMAGINING AND PLANNING INTERSTELLAR EXPLORATION

PAUL GILSTER

C

COPERNICUS BOOKS

An Imprint of Springer Science + Business Media

Published in the United States by Copernicus Books, an imprint of Springer Science+Business Media.

Copernicus Books
37 East 7th Street
New York, NY 10003
www.copernicusbooks.com

Book design by Jordan Rosenblum.

Library of Congress Cataloging-in-Publication Data
Gilster, Paul
Centauri dreams : imagining and planning interstellar exploration / Paul Gilster.
p. cm.
ISBN 0-387-00436-X (alk. paper)
1. Interstellar travel–Planning–Popular works.
2. Outer space–Exploration–Popular works. I. Title.
TL788.7G55 2004
629.43–dc22 2004056211

Manufactured in the United States of America.
Printed on acid-free paper.

9 8 7 6 5 4 3 2 1

ISBN 0-387-00436-X SPIN 10908815

FOR MY MOTHER

CONTENTS

PREFACE

I wrote this book because I wanted to learn more about interstellar flight. Not the Star Trek notion of tearing around the Galaxy in a huge spaceship—that was obviously beyond existing technology—but a more realistic mission. In 1989 I had videotaped Voyager 2's encounter with Neptune and watched the drama of robotic exploration over and over again. I started to wonder whether we could do something similar with Alpha Centauri, the nearest star to the Sun. Everyone seemed to agree that manned flight to the stars was out of the question, if not permanently then for the indefinitely foreseeable future. But surely we could do something with robotics. And if we could figure out a theoretical way to do it, how far were we from the actual technology that would make it happen? In other words, what was the state of our interstellar technology today, those concepts and systems that might translate into a Voyager to the stars?

Finding answers meant talking to people inside and outside of NASA. I was surprised to learn that there is a large literature of interstellar flight. Nobody knows for sure how to propel a spacecraft fast enough to make the interstellar crossing within a time scale that would fit the conventional idea of a mission, but there are candidate systems that are under active investigation. Some of this effort begins with small systems that we'll use near the Earth and later hope to extend to deep space missions. Indeed, NASA's interest in the solar sail as a propulsion system goes back to what might have been a rendezvous with Halley's Comet in the 1980s, a mission that never flew. The technology still excites scientists because solar sails use the power of sunlight, leaving heavy fuel tanks on the ground. We'll see the first solar sail flight experiments starting almost immediately. One of them, conducted by The Planetary Society of Pasadena, California, is even privately funded, perhaps a harbinger of a future with more commercial space ventures.

But I quickly learned that there are many interstellar options to consider, as well as many difficulties in evaluating them. Just how rigorous can you get with theoretical engines and hypothetical in-flight systems? Rigorous enough to produce a thoroughly researched study, the British Interplanetary Society's Project Daedalus in the 1970s, which was an attempt to design an unmanned probe to nearby Barnard's Star. Daedalus used nuclear fusion to power its way on a fifty-year journey. A type of fusion called inertial confinement fusion is now being considered as a way to sustain such a reaction and use it in an engine. The aborted Orion spacecraft, brainchild of Ted Taylor at General Atomic, was a design that set off nuclear weapons behind an ablative shield, protecting a manned spacecraft from their impact with a system of enormous shock absorbers. Remarkably, Orion might have taken a crew of a hundred to Saturn in the 1960s, but it was felled by the Nuclear Test Ban Treaty. Today we focus on controlled fusion and the possibility of igniting a sustained reaction by using tiny amounts of antimatter, a substance we can produce only in the billionths of grams.

But even nuclear methods are pushed to the limit by interstellar distances, which is why today's theorists continue to pursue antimatter hybrid systems driving uranium-enriched sails, and beamed laser propulsion for so-called lightsails, and exotic sail variations like magsails that get their push from the solar wind or a particle beam. Runways of fusion pellets may one day provide launching tracks for star-bound probes. Even Star Trek-style warp drives receive their share of study despite the odds against success.

A single insight from any one of a number of running experiments in different areas could change everything. The field is top heavy with ideas, alive with the loopy challenge of pushing a spacecraft to a sizable percentage of the speed of light. "Make no mistake—interstellar travel will always be difficult and expensive," said Robert Forward, "but it can no longer be considered impossible."

Forward, more than anyone, took interstellar missions from the realm of the imaginary into hard science. It was fitting that, in addition to his inestimable contributions to astronautics, he was also a science fiction author of great repute who figured out the parameters of a laser-pushed lightsail to the stars and incorporated them not only in scientific papers but in his novels. I set out to interview him as I began this book, only to learn of his death on September 21, 2002. Now I can only hope I have done justice to his many insights, and have in some way communicated my admiration for a career that created the serious study of things we always assumed were impossible. I also hope to catch a bit of his spirit, for science and science fiction weave a double strand throughout my text.

This book is written in homage to Forward and the many scientists and engineers who continue to pursue Centauri dreams. It is also a plea for patience. More than any other space project, an interstellar probe, whether to Alpha Centauri or any of a number of candidate stars, will demand a kind of long-term thinking that is now in short supply. The fastest mission we can conceive of still takes close to fifty years to get there and begin returning data. Many other assessments call for missions that last hundreds of years, and perhaps the most thoroughly analyzed concept using more conventional technology—what we could build in the near future—is fully a thousand years in duration. Do we have the vision as a species, or the patience, to wait out such a journey? In these pages I can only pose the question and hope to illustrate that we have done such things in the past. After all, deep changes in our attitude toward time seem a small thing to ask compared to the remarkable physics demanded by a journey to another star.

ACKNOWLEDGMENTS

This book would not have been possible without the cooperation and suggestions of many scientists who gave more of their time than this enthusiast had any right to expect. Special thanks go to physicist and author Gregory Matloff, whose willingness to talk about interstellar topics and to review the manuscript are much appreciated. At NASA's Glenn Research Center in Cleveland, I am particularly indebted to Marc Millis, head of the agency's Breakthrough Propulsion Physics project, who offered continuing support and ideas and reviewed part of the manuscript. Also at Glenn, physicist and science fiction writer Geoffrey Landis and physicist Edward Zampino were generous with their time, and Jean Schuerger was invaluable in following up on several key documents I needed.

At Marshall Space Flight Center in Huntsville, Alabama, NASA's John Cole, Sandy Montgomery, Randy Baggett, and Les Johnson offered insights into their own work, and a visit to Marshall's Propulsion Research Center gave me the chance to talk to Penn State's Raymond Lewis, a leading figure in antimatter research. None of these meetings would have taken place without the untiring help of NASA's Rick Smith at Marshall.

At the Jet Propulsion Laboratory in Pasadena, Humphrey "Hoppy" Price offered insights on solar sail technology, and I owe special thanks to Adrian Hooke, with whose help I attended a meeting of the InterPlanetary Internet working group. Thanks to JPL's Scott Burleigh and Leigh Torgerson, Robert Durst and Keith Scott of the MITRE Corporation, and Intel's Kevin Ford for the chance to sit in on that meeting, and especially to Vinton Cerf, inventor of the Internet's TCP/IP protocols, for extending the original invitation and answering questions. I also received valuable guidance from JPL's Moktar Salama and James Lesh on solar sails and interstellar communications respectively. Carolina Carnalla-Martinez helped me arrange the interviews at JPL and check facts. Steven D. Howe of the Los Alamos National Laboratories offered much help on nuclear issues and his own antimatter sail concept (Howe is now CEO of an antimatter company called Hbar Technologies). Robert Cassanova of NASA's Institute for Advanced Concepts provided the needed perspective on his agency's work.

Academic and commercial work on interstellar flight, particularly its propulsion conundrum, continues at a fast pace. At the University of Washington, Robert Winglee pursues the M2P2 magsail concept; thanks to him for a lengthy interview. Terry Kammash at the University of Michigan's Department of Nuclear Engineering and Radiological Sciences gave me the latest on Magnetically Insulated Confinement Fusion, and author and nanotechnology specialist Robert Freitas was generous with his time in describing how nanotechnology methods will make it possible to build probes as small as sewing needles. At Johns Hopkins University's Applied Physics Laboratory, Ralph McNutt pursues his "realistic interstellar probe" concept; his insights on fast missions with a close pass by the Sun for a gravity assist were much appreciated. Rensselaer Polytechnic Institute's Leik Myrabo explained how lasers may one day launch spacecraft into Earth orbit, while Embry-Riddle Aeronautical University's Laurence LaForge walked me through the issues involved in spacecraft autonomy.

Astrophysicist and consultant Jordin Kare explained fusion runways and his remarkable micro-sail concept, while James Benford of Microwave Sciences talked me through experiments demonstrating the effect of microwaves on sails. Gerald Nordley helped straighten me out on pellet propulsion and the needed "smarts" to keep the pellets on course. Princeton mathematician Edward Belbruno, who organized the landmark "Interstellar Robotic Flight: Are We Ready?" conference in New York in 1994, discussed that meeting in light of today's technology, while Bob Forward, Robert Forward's son, a screen writer in Los Angeles, gave me priceless insights into his father. James Woodward at California State University at Fullerton explained Mach's principle and its relevance to his own fascinating work on inertia and possible space drives, and the University of Colorado at Boulder's Webster Cash unlocked the mysteries of directly imaging other solar systems. Jordan Pollack at Brandeis's Dynamical and Evolutionary Machine Organization laboratory offered rich insights into machine intelligence and how it may evolve. Thanks, too, to Ann Druyan of Cosmos Studios and The Planetary Society for her thoughts on the Cosmos 1 sail and her reminiscences of her late husband, Carl Sagan. The chance to talk to Freeman Dyson, if only by telephone, fulfilled a lifelong ambition.

On a more personal level, my editor Paul Farrell at Copernicus Books offered encouragement when I needed it with his usual wry wit and a much appreciated sense of proportion. Lyman Lyons critiqued the manuscript word by word, offering insights that strengthened and illuminated the finished book. I appreciate his labors immensely. Special thanks go to my friends David Warlick, Jordan Smith, Barney Rickenbacker, and Kermit Ellis, each of whom encouraged me, offered suggestions, and knew just what questions to ask. Finally, my debt to my family is incalculable and growing.

CHAPTER 1

DESIGNING A MISSION

In a laboratory in the Propulsion Research Center at Marshall Space Flight Center in Huntsville, Alabama, a gleaming drum about the size of a whiskey barrel has been built that one day will hold a trillion of the world's rarest particles—antiprotons. The wry, kindly man standing next to me, wearing an unkempt shirt and battered khakis held up by suspenders, is Raymond A. Lewis, a key figure in developing Marshall's High Performance AntiProton Trap, or HiPAT. Lewis' eyes dance with the merriment of a man whose life's passion has taken him exactly where he wants to be. He is a physicist whose work on antimatter, both at Marshall and at the Swiss particle accelerator complex CERN (the French acronym for the European Laboratory for Particle Physics), has reshaped our understanding of how to trap and store antimatter.

Right now he's deadpan, explaining to this layman how he came to be laboring in an antimatter laboratory at NASA's key center for space propulsion. "I was working on collecting antiprotons in a trap," said Lewis, eyes twinkling. "And I thought, what can I do with them? They're too energetic to power a bicycle. They're too powerful to push a train. They'd overpower an ocean liner. And then it hit me. The only thing left was rockets!"

I had read Lewis's accounts of storing antimatter—as well as using it for deep-space propulsion—in the scientific literature, all written in the dry, precise cadences of physics. Now the man stood before me, unexpectedly gnome-like, a kindly uncle whose antic passion wasn't a stamp collection or a tool shop but the fabulous stuff that some people think may lead to the world's most powerful engine.

With a little imagination and a lot of time, antimatter might be the catalyst of fusion reactions that could propel instruments to the nearest stars, the Alpha Centauri triple star system, on a mission of perhaps a thousand years. And there was just the faint possibility that antimatter might, if ever produced in sufficient amounts, get a robotic probe there within the lifetime of the team that sent it. That idea is a long shot, to be sure, and antimatter's terrestrial use in medicine, positron emission tomography for the diagnosis of cancer and coronary disease, is far more immediate. But that doesn't stop NASA's gaze from turning skyward.

The reverse of normal matter, antimatter has opposite charge and spin, leading many to dub it "mirror matter." From a propulsion standpoint, an antimatter reaction carries a thousand times the energy of a nuclear fission reaction, a hundred times that of fusion. In rocket fuel terms, 1 gram of antimatter could produce the same energy as 20,000 tons of today's chemical fuel (a combination of liquid hydrogen and liquid oxygen). That energy is released when matter and antimatter meet, which is one reason an antimatter trap like HiPAT has to use extreme cooling techniques and strong magnetic fields to keep the antiprotons inside from colliding with the conventional matter of the trap itself.

Both Brookhaven National Laboratory in New York state and Fermi National Accelerator Laboratory in Illinois make antimatter. HiPAT will need to be taken to one of the labs to be loaded with antimatter and then brought home. Marshall intends to use a specially configured eighteen-wheeler to transport the system, including the central drum with its attached pipes and instruments. Antiprotons rolling down I-24 toward Huntsville may someday translate into fuel tanks for robotic probes pushing far beyond the solar system. Honk if you love the future.

I had come to Marshall on an unseasonably cool summer's day to learn more about the surprisingly large number of technologies that might take us to the outer planets, and perhaps one day to the stars.

- Antimatter in its various incarnations, as pure rocket fuel or, far more likely, as the catalyst for more conventional fission or fusion reactions.
- Solar sails and their cousins, sails driven by beamed laser light—let's call them lightsails—that use pressure from photons to drive vast film-like sails to a tenth of the speed of light.
- Ion drives that accelerate particles through an electrostatic grid to achieve hitherto unattainable exhaust speeds, creating engines that can stay in operation for years at a time (Marshall had recently shut down an ion engine that ran flawlessly for over five years on a test stand, a low-thrust device that has been demonstrated in space and which may open up the outer planets).
- Nuclear fission and fusion drives, especially the "nuclear pulse" concept once embodied in Project Orion, a proposal to propel a space vehicle by exploding atomic bombs behind a huge pusher plate, driving a payload that itself is cushioned by enormous shock absorbers. (Oddly, the idea seems sound, the crew's ability to survive being less of an issue than the intractable politics of nuclear explosions.)

No, the problem with interstellar flight is hardly a lack of ideas. They flourish in a variety of journals that are studded with extraordinary concepts. The thriving culture of scientific theorizing has led even beyond antimatter and lightsails to such exotica as the interstellar ramjet, a vehicle that scoops hydrogen from the interstellar medium itself to serve as fuel. This device reflects a larger principle: If you cut down the size and weight of the mission, you can get there faster and far more efficiently, which means find a way to leave some fuel behind, or extract it along the route.

Lately, there has been a flurry of interest in "pellet" propulsion, in one version of which streams of fuel pellets fired at the spacecraft would be ionized and reflected back as thrust by a magnetic field generated onboard the vehicle. Space "brakes" made of plasma that could use the solar wind from a star to decelerate a vehicle as it approaches the destination solar system have been examined, and so have fusion "runways," long chains of fusion fuel laid out along a tenth-of-a-light-year track to be made available to a starship as it accelerates, gobbling them up as it goes.

To promote new thinking on these and other ideas, NASA maintains its Institute for Advanced Concepts, a think tank based in Atlanta that welcomes convention-rattling theorizing—as long as it comes from sources outside the agency. Here and at NASA's Breakthrough Propulsion Physics project, the brainchild of Marc Millis at Glenn Research Center near Cleveland, the most arcane of theories are pondered to see if they measure up to serious scrutiny. Some of these are so theoretical that testing them seems impossible. The Alcubierre warp drive, for example, promises travel times to the stars not confined to the speed of light limit, while seemingly not violating Einsteinian relativity. That's a neat trick, and it may work, but early calculations by the Mexican physicist who gave his name to the theory showed that it would take all the energy in the universe to operate the drive (a later theorist cut that requirement down to the energy of one Sun-like star, an encouraging if still mind-boggling reduction in scale). Other concepts include inertial drives and "zero-point" fields that draw energy from the fabric of space itself. Even wormhole theory—entering a spacetime distortion and emerging perhaps half a galaxy away, or possibly in another time or even another universe—is fair game.

The impulse to push up against the boundaries of the plausible seems unquenchable among those who dream of the stars. For as Propulsion Research Center's John Cole told me at Marshall, conventional rocketry simply cannot do the job when it comes to interstellar distances. Asked by a schoolteacher how many Space Shuttles' worth of fuel it would take to reach Alpha Centauri, Cole's envelope-back mathematics yielded the result in kilograms of fuel. The figure was staggering: 10 to the 88th power, a number larger than the number of atoms in the universe. Chemical rockets are utterly inapplicable for the journey, but so are many of the ideas suggested as alternatives. We are left with an ongoing hunt for breakthroughs. "Many of the most intriguing of these ideas will fail," Cole said. "The odds on some of them are 99 percent against, maybe 99.999 percent against. But if we want to go to the stars, we have to keep looking because one of them may work, and we have no alternatives."

But on this summer day, as thunderstorms sprouted in the west, I was thinking about the past as much as the future. When

Henry the Navigator wanted to sail far down the coast of Africa for gold in the early fifteenth century, he faced a technology crisis. Square-sailed Portuguese ships could work their way down the African coast with the benefit of northerly winds, but they weren't suited for the return voyage. The triangular lateen sail, originally created by the Arabs, could work windward, but until this era had been used on craft too light to weather the Atlantic. Henry's shipbuilders combined existing Portuguese designs, heavy-hulled vessels, with lateen sails to create the caravel. Ready for deep-south missions, Portuguese sailors used the quadrant and astrolabe for navigation and, unable to get a fix on Polaris in southern waters, employed a newly formulated "rule of the sun" to help determine latitude.

One technology invariably points to the next. Imagine the world fixated on its television screens as the first images from an Alpha Centauri probe begin to reach the Earth. Is such a Voyager-style mission to another star possible? The answer is yes, with a lot of qualifications, and with the same kind of technological bootstrapping Henry used to push ever south.

Because Alpha Centauri is 4.3 light years away, it takes a radio signal 4.3 years to return from there, just as long as it takes light. So forget about controlling a Centauri probe from the Earth. The best you can hope for is to make the occasional navigational correction, and this only if the craft is traveling at the relatively slow (in Einsteinian terms) speed of one-tenth the speed of light. A laser beam pushing a lightsail could also contain data, and the beam itself might be useful in orienting the probe—stay on the beam and the probe gets to its destination.

But all malfunctioning systems—all valves that don't close and data devices that jam—will have to be repaired on board. An interstellar probe will include robotic systems of a sophistication far beyond anything we have built before. It will probably be self-aware in some fashion and will make adjustments depending on what it finds circling the Centauri stars. It will use laser signals to communicate with the Earth, returning data to a vast, solar system-wide network of telescopes, computers, and radio and optical receivers that will be able to pluck its infinitely tiny signals out of the starshine. It will be made of components that will function fifty years, or a hundred, or perhaps much longer, because systems cannot be allowed to fail four light years from home.

There are reasons why a tenth of light speed, or .10c (*c* is apparently from the Latin *celeritas*, "swiftness," and is the symbol for the speed of light), is something of a Holy Grail, although there are some designs, like Robert Forward's Starwisp, that could outpace even that. But give us ten percent of light speed and we can do a star mission. Imagine a fifty-year journey and the support that will be needed to keep it going. NASA scientists like .10c because it keeps such a mission within the lifetime of a researcher, and doubtless there will be some on the Centauri team who will see both ends of the journey. But most of the scientists who will complete the mission will inherit their work. An interstellar probe will be handed off from one generation to another, and its mission presupposes that we will have the determination to see it through. That is no small task in an era of quarterly-return thinking, when all that seems to matter is what appears on the next bottom line, and the jittery, self-gratifying rhythms of post-modern culture, emphasizing fragmentation and discontinuity, dominate the public arena. Where now is our notion of the long haul?

In a way, a Centauri probe isn't modern. Paradoxically, it may require medieval thinking, the sort of thinking that built cathedrals in Chartres and Salisbury and Cologne. Just as the stonecutters and masons of twelfth century France found themselves awash in a sea of endless time, dwarfed no less by the monuments they built than by the vastness of the eternity they sought to suggest in their stained glass windows, so the builders of the first star probe will be overwhelmed by the immensity of distance.

There are various ways of looking at the distance problem, but consider first the Voyager spacecraft that gave us such spectacular images of the outer planets, culminating in Voyager 2's encounter with Neptune in August of 1989. Voyager 2 is now leaving the solar system behind at a rate of 3.3 astronomical units (AU) every year. An AU is the mean distance between the Earth and the Sun, used as a yardstick for distances within the solar system. At approximately 150 million kilometers, or 93 million miles, an AU covers a lot of territory, but Neptune is thirty times that distance from the Sun.

Voyager 1 is the fastest manmade object. Traveling at 3.6 AU per year, it has reached a speed that would take it from New York to Los Angeles in less than four minutes. Even at that rate, it would take Voyager 1 over seventy-seven thousand years to reach Alpha Centauri. If we could send an interstellar probe at 100 times the speed of Voyager 1, we could reach Alpha Centauri in less than a millennium. Is an almost 800-year mission possible, and would we have the political, financial, and scientific will as a species to follow it through to its destination? Or would we wait until .10c is within our reach? These are the kinds of questions that tantalize and frustrate researchers, and their answers—as well as their solutions to the intractable propulsion problem—are all over the map. The one constant is the vast gulf any interstellar probe will confront, a void so deep that while its numbers are routine to specialists, they are constantly restated at conferences and in the technical literature, as if to remind the cognoscenti that the proper response to even nearby star travel is something akin to awe.

The English astronomer Sir John Herschel (1792–1871), a wizard at the study of double stars who was the first to try to measure the distance to Alpha Centauri, liked to imagine stellar distances in terms of the transportation systems of his era. Thus ". . . to drop a pea at the end of every mile of a voyage on a limitless ocean to the nearest fixed star, would require a fleet of 10,000 ships of 600 tons burthen, each starting with a full cargo of peas." If vegetables don't suffice, we can use fruit to visualize interstellar distances. If we reduce the size of the solar system by a factor of one billion, we wind up with a Sun that is roughly 1.5 meters (4.9 feet) in diameter. A city block away is the Earth, about the size of a grape, while the grapefruit-sized Jupiter orbits five blocks away, and distant Neptune is fully thirty blocks out. In these terms, with humans reduced to the size of atoms, Alpha Centauri is 40,000 kilometers (almost 25,000 miles away). In AU terms, we must traverse 270,000 times the distance between the Earth and the Sun to reach the primary Centauri stars; Proxima Centauri is slightly closer at 260,000 AU.

Two hundred and seventy thousand times the distance between the Earth and the Sun is roughly 4.3 light years, meaning a beam of light would travel for 4.3 years before reaching the goal. Amir D. Aczel, an author and mathematician, likes to think in terms of jet planes. Traveling at some 600 miles per hour, a jet would travel 5.25 million miles per year, which means that you would be on a jet for six years to reach Mars at its closest to Earth. The distance to the Sun is roughly seventeen "jet years." On that scale, you would fly on a jet almost five million years to reach Alpha Centauri (not to mention reaping plenty of airline miles).

And maybe this analogy, created by the Jet Propulsion Laboratory's Rich Terrile, will give an idea about the average distances between the stars. Take a box of salt and pour it out on a table. You're looking at about five million grains of salt; two hundred such boxes make a billion grains. Twenty thousand boxes make one hundred billion grains, roughly the number of stars in the Milky Way Galaxy. If you spread the salt out to scale to mimic the actual separation of the stars in our vicinity of the Galaxy, the salt grains would each be seven miles apart.

The problem with analogies is that a little familiarity goes a long way. Who doesn't remember running into such numbers in grade school or high school, playing the same game: "Imagine the Sun is a beach ball. Now, how far away is the Earth?" Like the scaling problem teachers have when explaining atoms and molecules, the stellar distance problem defies our understanding so utterly that we dismiss it, knowing only that the stars are far away indeed. They are also spatially challenging because we don't always visualize them as existing in a three dimensional universe, even though we know intellectually it is true.

But go to Barnard's Star, at 5.9 light years one of the closest to our solar system, and look up into the night sky from a hypothetical planet there. Almost everything you see will be familiar, because on a galaxy-wide scale, such a voyage is tiny. Most of the constellations look the same except for Orion, whose famous belt now shows not three but four stars. (The fourth star is our Sun.) Move in a radically different direction to Alpha Centauri. Stopping on a planet there, you would see the same constellations, only here Orion looks just as it does from Earth. Now it is Cassiopeia that draws our attention. The mother of Andromeda and wife of Cepheus, her stars form a celestial W or M depending upon your orientation. In a view from Alpha Centauri, Cassiopeia's five stars are joined by a sixth, a first magnitude beacon that is our Sun, while the constellation Centaurus is now absent its brightest component, our vantage point Alpha Centauri. Interstellar flight demands that we navigate such seas, awash in immensity, trillions of miles from home, in a place where our own Sun is just another star.

Scientists who run into impossible problems have an uncanny knack for solving them. Often they gather in conferences to debate the possibilities, as they did in July of 1998 at the California Institute of Technology. Sponsored by the Advanced Concepts Office of the Jet Propulsion Laboratory and NASA's Office of Space Science, the conference bore the provocative title "Robotic Interstellar Exploration in the Next Century." Dr. Steven D. Howe, a Los Alamos-based physicist, may have caught the spirit of the meeting better than anyone when he showed a slide saying, "Any technology being presented to you at this workshop will need at least one miracle in development in order to enable an interstellar mission." Howe went on to describe why the word "miracle" is applicable. A ten-year mission to the Kuiper Belt, an unexplored band of cometary objects at the edges of the solar system, requires a velocity of 117 kilometers per second. To reach the Oort Cloud—a halo of perhaps trillions of comets 10,000 times as distant as the Sun from the Earth—in forty years demands 1,200 kilometers per second. An Alpha Centauri mission in a forty-year time frame would call for a mind-numbing 30,000 kilometers per second. Only the first of these speeds is within range of present-day technologies.

The purpose of the Caltech conference was to explore the possibilities of sending an unmanned craft into interstellar space. Not all such thinking is devoted to nearby stars, of course, for a mission is "interstellar" if it pushes into the interstellar medium, defined as the area beyond the heliosphere, the "bubble" in space produced by the solar wind. Because almost all the material within the heliosphere comes from the Sun, scientists have great interest in learning more about the interstellar medium beyond it, which should tell us much about the Galaxy and how it formed. The Kuiper Belt and the Oort Cloud are primary targets. And there is speculation that a spacecraft placed at the right distance from the Sun (about 550 AU, beyond the heliosphere) might use the gravitational lensing effect predicted by Einstein to see distant parts of the universe.

In terms of distance, even the Kuiper Belt offers a huge challenge. Pluto, the last known planet unvisited by spacecraft, orbits at an average 40 astronomical units (AU) from the Sun. First predicted by the Irish amateur astronomer Kenneth Edgeworth in 1943, the Kuiper Belt was nonetheless named for American astronomer Gerard P. Kuiper, who proposed in 1951 that there must be a band of comets close enough to account for those that, like Halley's Comet, appear on a regular basis (i.e., within short enough time frames to be observable and confirmed by astronomers, a span normally defined as two hundred years). Kuiper also noted that such comets emerge into view within the ecliptic, the plane within which all the planets except Pluto orbit. Kuiper thus deduced a broad belt of comets, cometary debris, and icy asteroids beginning inside the orbit of Pluto and extending to approximately 1,000 AU.

The first Kuiper Belt object was discovered in 1992, and numerous others have since been documented, including large planetoids like Sedna, which is fully three-quarters the size of Pluto. Gravitational interactions between the Kuiper Belt comets may disturb their orbit sufficiently to allow them to be pulled into the inner solar system. There may be hundreds of millions of comets in the Kuiper Belt; in fact, astronomer John Davies, a specialist in small solar system objects, estimates that there are 70,000 objects larger than 100 kilometers in diameter in the region between 30 and 50 AU, existing in a surprisingly thick band. Many regard Pluto as merely the largest Kuiper Belt object, with the Neptunian satellites Triton and Nereid as possible Kuiper inhabitants that were captured long ago by the gravity of their planet.

Thus we have two potential targets for interstellar missions just outside the solar system. For beyond the Kuiper Belt is the Oort Cloud, named after Dutch astronomer Jan Hendrik Oort. Oort's work in 1950 convinced him that a huge cloud of comets existed well beyond the outer planets, a cloud that would be invisible to astronomers because we have no instruments sensitive enough to detect it. Oort knew that no comet has been seen to approach the Sun with an orbit implying it was unrelated to the solar system. In addition, he had noticed that the orbits of long-period comets seemed to take them as much as 50,000 AU from the Sun. His deduction of a vast system of perhaps one trillion comets followed, a cloud that may boast more mass than the planet Jupiter. Oort's Cloud is conceived as a spherical cluster centered on the Sun that could be as much as one light year in diameter; cometary orbits in such a cloud could take millions of years to complete. Whereas Oort comets seem to have no preferential direction of arrival, the much closer Kuiper Belt comets emerge within the orbital plane of the planets, indicating the cloud-like shape of the former and the flattened shape of the latter.

The Caltech conference would not be the first or last conference to explore interstellar missions. The results of their work will infuse these pages. Subsequent meetings would focus more tightly on reaching nearby stars like Alpha Centauri or Epsilon Eridani. A landmark of sorts was a gathering held at New York University in 1994. Called "Practical Robotic Interstellar Flight: Are We Ready?", the conference was sponsored by The Planetary Society (with help from NASA and the Jet Propulsion Laboratory) and organized by mathematician Edward Belbruno, a wizard at chaos theory and the calculation of exotic orbits. Issues addressed at the university sessions (along with a meeting at the United Nations building) ranged from the choice of target stars to the seemingly staggering costs of such a mission. The theorizing was hot and heavy, and good papers resulted, published later in the *Journal of the British Interplanetary Society*. But interstellar theory remained more of an avocation than a serious research goal with NASA until four years later, when everything changed.

On July 3, 1997, as the Pathfinder spacecraft neared its touchdown on Mars, NASA administrator Daniel S. Goldin told agency science chief Wes Huntress and assembled reporters that building a robotic probe that could reach another star would now be a goal for NASA. "We have to set goals so tough it hurts—that it drives technology—in semiconductors, materials, simulation, propulsion."

NASA had already examined two interstellar mission concepts in the 1970s and 1980s. The first of these was known loosely as the Interstellar Precursor Mission, a study developed at the Jet Propulsion Laboratory in 1977. This hypothetical mission was designed to explore the heliopause—where the influence of the charged particles streaming out from the Sun effectively ends—and the nature of the interstellar medium, measuring the distribution of interplanetary gas and studying Pluto and its moon Charon along the way. With a primary mission duration of twenty years (and an additional fifty years of what NASA calls an "extended mission" after the primary objectives have been fulfilled), the Interstellar Precursor Mission would be designed to penetrate deep space out to perhaps 1,000 AU.

A second concept was a modified and modernized version of the first. It would come to be known as the Thousand Astronomical Unit (TAU) mission. Here the primary goal was to take the Galaxy's measure, studying its stars through so-called stellar parallax astrometry, which mines data from a distant spacecraft and compares them to the view from Earth to develop more precise readings on stellar motion and distance. TAU was also to be a platform for other astrophysics and astronomy projects, examining the heliopause, the interstellar medium beyond it, and, if possible, the planet Pluto. Using advanced optical communications to transmit its findings, the 1,200-kilogram probe would travel 1,000 AU in a fifty-year flight. Both Interstellar Precursor and TAU were designed around nuclear-electric propulsion systems, the nuclear version of a technology that would later be successfully tested aboard the Deep Space 1 spacecraft. Solar sails unfurled close to the Sun also remained an option.

But it is one thing to design hypothetical missions, quite another to be given an exhortation from the top to make them happen. Taken off guard, many NASA scientists embraced Goldin's goal while acknowledging the immensity of the task. As research began to come together, the agency formed the Interstellar Probe Science and Technology Definition Team (IPSTDT) that assembled at the Jet Propulsion Laboratory in Pasadena for a preliminary meeting in February of 1999. Later that year, Goldin made it clear that the stars might require entirely new thinking. In a presentation to the 100th anniversary meeting of the American Astronomical Society in June, the administrator described lightsail technologies, advances in neural nets, genetic algorithms that "learn" as they process code, and hybrid systems that fuse biology with silicon.

All these breakthroughs could be employed in an interstellar probe he envisioned as a "Coke can-sized spacecraft" that essentially builds itself into a star probe using carbon, iron, and other materials on a nearby asteroid. "This reconfigurable hybrid system can adapt form and function to deal with changes and unanticipated problems," said Goldin. "Eventually it will leave its host carrier and travel at a good fraction of the speed of light out to the stars and other solar systems.... Such a spacecraft sounds like an ambitious dream, but it could be possible if we effectively utilize hybridized technologies."

Among the papers of physicist Robert Forward, maintained after his death by Salmon Library at the University of Alabama in Huntsville, are electronic mail printouts of the scheduling and decision making that went into the Interstellar Probe Science and Technology Definition Team. The team, wrote the Jet Propulsion Laboratory's Richard Mewaldt, had been charged with defining the first mission to explore the local interstellar medium, reaching a distance some 200 AU from the Sun within fifteen years by using a solar sail for propulsion. The JPL probe, as outlined in a concept briefing Forward delivered to the IPSTDT at Beckman Institute Auditorium at the California Institute of Technology on February 16, 1999, would be designed to explore the structure of the heliosphere and its interaction with the interstellar medium, while studying the astrophysical processes occurring in both, all of which would have implications for our understanding of the origin and evolution of matter in the Galaxy. Such a mission would be the first time scientists could measure the plasma, neutral atoms, magnetic fields, dust, energetic particles, cosmic rays, and infrared emissions flowing from our solar system into surrounding space. This would answer many questions about how the Sun interacts with the rest of the Galaxy as well as how matter is distributed throughout the outer solar system.

Hired to investigate the feasibility of candidate propulsion systems for the mission, Forward seemed the ideal man for the job. A physicist, aerospace engineer, inventor, and popular science fiction author, he had pioneered futuristic notions from antimatter to gravitational engineering, space tethers, and rocketless propulsion. In 1975 he was asked by the House Committee on Science and Technology to come up with a national space program for interstellar exploration, the first time to my knowledge that such things as fusion rockets and laser-pushed lightsails have appeared in the Congressional Record. Audacity defined the man, as witness these remarks to the committee, which are worth quoting at some length:

> A national space program for interstellar exploration is proposed. The program envisions the launch of automated interstellar probes to nearby stellar systems around the turn of the century, with manned exploration commencing 25 years later. The program starts with a 15-year period of mission definition studies, automated probe payload definition studies, and development efforts on critical technology areas. The funding required during this initial phase of the program would be a few million dollars a year. As the automated probe design is finalized, work on the design and feasibility testing of ultra-high velocity propulsion systems would be initiated. Five possibilities for interstellar propulsion systems are discussed that are based on 10- to 30-year projections of present-day technology development programs in controlled nuclear fusion, elementary particle physics, high-power lasers, and thermonuclear explosives. Annual funding for this phase of the program would climb into the multibillion-dollar level to peak around 2000 A.D. with the launch of a number of automated interstellar probes to carry out an initial exploration of the nearest stellar systems. Development of man-rated propulsion systems would continue for 20 years while awaiting the return of the automated probe data. Assuming positive returns from the probes, a manned exploration starship would be launched in 2025 A.D., arriving at Alpha Centauri 10 to 20 years later.

All this at a time when the war in Vietnam had just ended and the United States was in full retreat from the Moon. If an idea had the stamp of future breakthrough on it, chances were Forward had already written a paper on it. And when it came to the kind of solar sail that the interstellar probe team hoped to use to reach out 200 AU, Forward had established a handsome list of publications on the concept, including the first serious study of how a beamed lightsail could not only be sent to another star, but could be decelerated upon arrival by the laser beam that pushed it.

Flamboyant, given to loud vests made for him by his wife, highly visible with his shock of white hair, Forward would become a fixture on the interstellar research scene, not to mention the numerous public meetings and science fiction conventions he attended. A scientist to the core, he saw his writing as a separate and distinct occupation. "The reason I write is to teach," Forward wrote in an unfinished autobiography as cancer consumed him in late 2002. "Every time I found a facet of science that I thought was interesting, I would try to pass on that information to the public." This he would do in sixty-one popular science articles, eleven science fiction novels, and twenty-one short stories, not to mention the one hundred thirty-four publications on aerospace topics, thirty-five papers in physics, and twenty-eight patents that defined his career.

Forward's presentation on the JPL interstellar probe to the team assembled at the California Institute of Technology that day in early 1999 was a stunner. Having outlined the mission's science objectives and constructed a hypothetical payload of instruments, he went on to provide a solar sail technology roadmap that began with a sail demonstration in an upcoming mission called Geostorm, which would use a sail to position a payload that would provide early warning of solar storms for military and civilian satellite users. Forward envisioned later science missions within the solar system ranging from sampling a comet to a spacecraft designed to land on Jupiter's moon Europa. What he called "interstellar medium exploration" used a solar sail to power the JPL mission under discussion, but he saw future missions to the Oort Cloud using a lightsail pushed by a beam of laser power. The culmination would be the first true interstellar probe, a beamed lightsail four kilometers in diameter that would make a flyby of the Alpha Centauri system in forty years. Such interstellar sails might one day be expanded to a behemoth 1,000-kilometer sail that he believed could reach stars up to forty light years away.

The optimistic dates in Forward's presentation have already slipped (he had spoken of sail demonstrations by 2002, with launch of the heliosphere explorer by 2008, a schedule now obviously impossible). One reason, of course, was to be found on the last line of the last slide in his briefing, one that read "Cost: TBD." TBD means "to be determined," a phrase that covers a lot of ground for agencies that rely on public policymakers for their funding. But the first solar sail experiments in flight are upon us, with the development of the Planetary Society's Cosmos 1 sail and another privately funded sail by a Texas group called Team Encounter. NASA's own plans call for flight testing of solar sails perhaps beginning as early as 2008, and the European Space Agency is also investigating sail technologies. No firm date on the development of JPL's interstellar probe can be given, but it is clear that the first spacecraft to use interstellar propulsion technologies will be propelled either by a sail or a form of nuclear electric engine.

The work within an American space agency shouldn't disguise the fact that interstellar research is a truly international phenomenon. The European Space Agency's solar sail work, led by Scottish scientist Colin R. McInnes, is advanced, as we'll see in chapter 5. For that matter, the solar sail concept Robert Forward worked with that day at Caltech also drew heavily on the work of Italian physicist Claudio Maccone and the American theorist Gregory Matloff, like Forward a proponent of interstellar flight whose credentials are well established in the field. Maccone and Matloff had studied sail designs for missions out to the 550 AU point, the zone where the Sun's gravity acts like a lens that can be used to magnify distant astronomical objects. Matloff published an analysis of design options for such a probe, using a 10,000 square meter sail as his baseline, and calculated a flight time of sixty years to the gravity focus. Italian physicist Giovanni Vulpetti would draw on Maccone and Matloff's mission concepts to create a more conservative solar sail mission to the heliopause called Project Aurora.

A sequence of future events began to take shape for interstellar theorists, although its dates could only be guesswork. A mission to penetrate the heliopause would go first, followed by a probe designed to study the Kuiper Belt. The Oort Cloud was a goal for a later probe, one that would probably not fly until mid-century. The ultimate goal, Alpha Centauri, hovered like a beacon beyond the Oort Cloud mission. It was becoming clear that

NASA would not be able to launch a Centauri probe without serious foundation building.

But that foundation has been under construction for a long time, and no one has worked at it with more eloquence or insight than Gregory Matloff. He brings not simply enthusiasm but genuine commitment to the notion that humanity's future is in the stars. And over the past three decades, he has emerged as a key figure in interstellar studies. Matloff found his subject when, as a graduate student at New York University, he read James Strong's 1965 book on interstellar probes, *Flight to the Stars*. Having found an error in Strong's discussion of fusion, he wrote a paper comparing different methods of using excess fusion energy. Finishing a PhD on planetary atmospheres at NYU, Matloff was soon presenting papers on concepts that were unashamedly futuristic. This was the 1970s, a time when the line between science and science fiction seemed to be defined by Gerard O'Neill, who envisioned enormous colonies at the gravitationally stable Lagrangian points in nearby planetary space. Matloff would imagine O'Neill's colony designs wedded to propulsion systems of the sort envisioned by Project Orion and a later design called Daedalus.

The result was a "worldship," an interstellar ark that would carry colonists on a thousand-year journey to another star, using a form of electromagnetic braking (of the sort later refined by Robert Zubrin and Dana Andrews) to decelerate upon arrival. The idea was straight out of Robert Heinlein, although the worldship Heinlein envisaged in his 1941 story "Universe" (later reshaped as the novel *Orphans of the Sky*) was nothing like an O'Neill colony, with its vast farmlands, rivers, and almost pastoral ecology. Brian Aldiss would work this theme in far different ways in his 1958 novel *Non-Stop* (published in the United States as *Starship*), but both stories played off the idea of a multi-generational voyage whose crews have gradually lost the memory of their mission. Matloff's daring was to find plausible scientific ways of getting a worldship to its destination, demonstrating that in terms of design, a starship could be extrapolated from today's physics.

But interstellar studies is a restless field. Concepts mutate and re-emerge, old notions take on vibrant new forms. And while fusion seemed to be stuck in the endless attempt to reach the breakeven point, beyond which more energy comes out than goes into making the reaction, the hot technology of the late 1970's was the solar sail. Proposed by a team at the Jet Propulsion Laboratory for NASA's never-flown Halley's Comet mission, sail technologies were a more ingenious and considerably more plausible way to accelerate Matloff's worldship. The physicist now took upon himself the challenge of designing a thousand-year journey that would reach Alpha Centauri pushed by nothing more than sunlight. To gain the needed speed, Matloff's starship would perform what science fiction writer and fellow physicist Gregory Benford came to call a "sun-diver" maneuver, unfurling its sail as the vessel all but skimmed the Sun's corona before heading out of the solar system.

Matloff's analysis of such missions made his participation in any interstellar spaceship design a priority. Contacted by Robert Forward about doing work on the NASA project, Matloff joined the team as a consultant. Active as ever, he now looks back on that effort with a certain wistfulness. "Back in 1999 and 2000, we thought, naively enough, that we could go directly to a star," Matloff told me in a recent interview. "Now we realize we have to put solar sails through a period of near-Earth applications first. We really believed we could skip a lot of that. Which of course was naive. We came to realize that we have to test out the concept much closer to home."

This is more or less how things stood in early 2002, when I became attracted once again by the interstellar conundrum. Goldin had recently resigned from his post at NASA, and while the agency's theorists appreciated his enthusiasm for the task, a quick push for Alpha Centauri was clearly not in the cards. I wanted to learn what had happened to the interstellar idea. Had it simply been abandoned, or modified so extremely as to be unrecognizable? Was an interstellar probe even possible? This book is the result of my attempts to find answers. It is also something of a meditation about the nature of interstellar travel and how it plays around in our society's subconscious, surfacing in the form of vivid science fiction imagery that, even when inaccurate or outmoded, still transmits a passion for the enterprise that has driven otherwise sober scientists to dream of the stars.

I had been a reader of science fiction since boyhood, attracted by magazines like *Astounding Stories* (later *Analog Science Fiction and Fact*), edited by the canny John Campbell. I had read H. L. Gold's *Galaxy* with pleasure as it pushed past overt science into the mechanics of future societies, probing their assumptions, looking at what strange outcomes do to human relationships. In these and other magazines, many of whose space-themed covers took my breath away, starflight was more or less a given. It had been so since 1928, when Edward E. Smith, a Wisconsin-born chemist who specialized in doughnut mixes, took rockets out of interplanetary orbits and opened up the universe in his *Skylark of Space*. And in the way of science fiction readers, I had developed a healthy dualism. There was the real world, and there was the future—or sometimes parallel—world of science fiction. As I grew older, I never lost either my fascination with starships, or my conviction that while we would one day have them, their technology was so far beyond our own as to be, in Arthur Clarke's fine phrase, "indistinguishable from magic."

Imagine my surprise, then, to reach my sixth decade with assumptions intact, only to realize that there was a flourishing starship culture that found its inspiration in pure science. Its physicists, astronomers, and engineers worked in the pages not of science fiction magazines but learned journals with titles like *Acta Astronautica*, *The Journal of Spacecraft and Rockets*, *Icarus*, and even the prestigious *Physical Review*. They seemed to share an assumption with Robert Forward, who wrote in 1996 that ". . . interstellar travel will always be difficult and expensive, but it can no longer be considered impossible." Not even, Forward hastened to add, with today's technology extrapolated forward no more than a few decades.

Just as science fiction had its classics of starflight, like Poul Anderson's dazzling tour of the universe in his 1970 novel *Tau Zero*, so this other culture, poised between science fiction and physics, had developed its own literature studded with keepers.

Forward's own "Round-Trip Interstellar Travel Using Laser-Pushed Lightsails," which appeared in the *Journal of Spacecraft and Rockets* in 1984, was only one of a vast bibliography of scientific papers whose imagination was matched only by their rigorous methodology. Many of these appeared in the *Journal of the British Interplanetary Society*, the publication of an influential group of space enthusiasts whose work before and after World War II brought public attention and professional respect to space travel. (Arthur C. Clarke was a member, and it was to the British Interplanetary Society that he would first present many of his ideas.) *JBIS* has become the key interstellar periodical; the titles of its articles out-do anything John Campbell ever published in *Astounding*, and it was the BIS that produced the first full-length study of a robotic interstellar probe, the fusion-driven Project Daedalus.

Who could fail to be struck by titles like A. R. Martin's "Project Daedalus: The Ranking of Nearby Stellar Systems for Exploration," or R. W. Stimets and E. Sheldon's "The Celestial View from a Relativistic Starship"? How about Gregory Matloff and Eugene Mallove's "Solar Sail Starships: Clipper Ships of the Galaxy"? The literature seemed inexhaustible. Robert Forward and astronautical engineer Eugene Mallove, who published several bibliographies of interstellar studies in *JBIS*, were eventually overwhelmed by the page count and forced to abandon the effort—their 1980 bibliography included fully 2,700 items in seventy subject categories. Each paper led to others: I found studies in profusion on propulsion, communication systems, interstellar navigation, shields to protect against collisions with interstellar dust—all these explored every facet of building and flying a probe to nearby stars. What was afoot was foundation building. When we do one day go interstellar, we will draw from these papers the basic principles that will guide the mission.

I also discovered there was a parallel story, one that starts not with extrapolation and theory but current technology. We have been building space probes for a long time, long enough to know what sort of planning and experiment must go into every design. An Alpha Centauri probe, whether it flies in fifty years or a hundred, will grow organically out of concepts that are being tested in laboratories and in some cases flown on missions today.

NASA's current parallel story can be best seen in a vehicle called Deep Space 1. Launched in October of 1998, the spacecraft was designed to test twelve advanced, high-risk technologies including a solar/electric ion propulsion system that accelerated continuously, providing a ten times better ratio of thrust to propellant used than a chemical rocket. Deep Space 1 also carried an optical navigation system that gave promise of autonomous operations, allowing a distant probe to compute and correct its own course rather than waiting on signals that might take days, or much longer, to reach the vehicle from Earth. Add a raft of other tests, including status monitoring systems for on-board condition reports, concentrated solar arrays that cost less but generate power more efficiently than conventional systems, and "agent" software that is designed to allow a spacecraft to operate without human intervention, and you have a design for the future. When it was retired from its extended mission in late 2001, Deep Space 1 had returned images of Comet Borrelly that were among the best ever seen of a comet, and had successfully completed technology tests of hitherto unproven flight systems.

Managed by the Jet Propulsion Laboratory, Deep Space 1 was the first of NASA's so-called New Millennium missions, a program designed to test and validate new technologies so they could be used in future missions without risk. And risk is a big word inside NASA. No one wants to sign off on a mission that fails; managers are rewarded for success, not for risk-taking. The result is that the agency has often been criticized for using components with a long history, and putting more emphasis on system redundancy than innovation. Thus New Millennium, the brainchild of former NASA administrator for space science Wes Huntress. Some of the technologies the Jet Propulsion Laboratory managed on Deep Space 1 may lead to key components for interstellar probes.

You would think that if there were a place where science met science fiction, it would be at the Jet Propulsion Laboratory, where so much of NASA's work on robotic space exploration proceeds. But while many a space scientist admits to reading science fiction on the side, there is frequently an invisible barrier between the two activities. We protect those things that are closest to us, and it may be that working on, say, a Mars rover makes it more likely that you will be less willing to have your work taken lightly, as mere fodder for science fiction buffs. In any case, wild speculation has little place in the realm of Voyager, Galileo, and Cassini, whose missions proceed with precision measurement, quantified data, and an uncanny touch for squeezing information out of distant signals.

In his sunlit office at JPL one morning in late January of 2003, Humphrey Price talks about our future in space. He is an earnest, friendly man whose passion for ideas drives his rapid-fire speech. As NASA's solar sail lead investigator, Price is universally known by his nickname of "Hoppy." "You'll have to ask Hoppy about that," scientists would tell me, assuming I knew the reference. Microwave physicist James Benford, discussing methods of solar sail deployment in space, told me that he favors spinning the sail to ensure an even distribution of its material. "It's a method I like, and so does Hoppy," Benford said. And it becomes clear that when discussing issues involving NASA in deep space, Hoppy Price is a man to be reckoned with.

Deep space exploration in the next forty years will almost certainly involve both solar sails and nuclear electric engines, even as more exotic possibilities are tested in the laboratories. NASA's In-Space Propulsion Technology Project, which evolved directly from the propulsion research effort within the Interstellar Probe Science and Technology Definition Team, is managed out of Marshall Space Flight Center in Huntsville, but the Jet Propulsion Laboratory is also a major center for solar sail activity. Price's time is spent looking at upcoming missions that might use sail technology, missions like Geostorm, designed to provide early warning of solar flares by using a solar sail to approach to no more than a fifth of the distance that the Earth is from the sun, using photon pressure to balance the sun's gravity. Or the Solar Polar Imager, a sail mission that approaches the sun from a spiraling, equatorial orbit that is then altered to push the spacecraft into a circular polar orbit perpendicular to the ecliptic. No currently available ion drive or chemical rocket could achieve that kind of orbit because of the large change in velocity required.

But while he's looking out for the near term, Price is also an advocate for long-term thinking. Acknowledging the need for a space-based infrastructure to support a Centauri mission, he thinks the time frame of the mission is an equal challenge. "Robert Forward talked about getting there in fifty years or less, a time scale that seemed to make sense because it would equal the possible lifetime involvement of a researcher," Price said. "What may be more reasonable is to take a little more time. Because we're also working on the beginnings of a program to build very long lifetime electronics, systems that can operate for up to two hundred years. If you let yourself take two, even three hundred years to get there, the problem of propulsion becomes a bit easier. We as a culture may have to start thinking in terms like that. The average worker on a medieval cathedral didn't live to see it completed. My view is that the first time we send something to Alpha Centauri, it will probably take hundreds of years to get there."

Hundreds of years or the lifetime of a researcher? A robotic interstellar probe challenges our commitment to the future. One interstellar mission concept, conceived by a man who consulted on the JPL interstellar probe project, backs up to view human science within a millennial frame. At Johns Hopkins, physicist Ralph McNutt has analyzed strategies for an interstellar precursor mission that would both explore the near interstellar medium and set the stage for future interstellar operations. McNutt's work, prepared for NASA's Institute for Advanced Concepts, envisions a heavily shielded spacecraft using a gravity assist from the Sun (after first being assisted by an initial flyby of Jupiter to facilitate the near-Sun approach), closing to within three solar radii. At this point, the probe is given an additional kick by the burn of a solar-thermal rocket (using the sun's heat to energize a gas propellant, or conceivably using a scaled-down "Orion" approach with nuclear explosions behind the craft to push it), then separating from the propulsion system to begin its flight toward the interstellar medium. With such methods, McNutt's probe reaches a speed of 20 AU per year, with the goal of 1,000 AU from the Sun in less than 50 years. McNutt's probe would pioneer the technologies required for long-term autonomous spaceflight, paving the way for a mission to Epsilon Eridani late in the twenty-first century.

How to fit an interstellar probe into the human experience? The monuments of antiquity—the Great Pyramid of Cheops, the cathedral of Chartres, or the Shinto temple complex in Japan known as the Ise Shrine—consumed the work of generations. Indeed, the Ise Shrine still does, its all-wood structure being carefully rebuilt every twenty years for the last thousand. One way to view an interstellar robotic probe is as a cross between a scientific mission and a gift to the human future.

Thus McNutt writes about a second generation interstellar probe traveling at ten times the speed of the precursor. At 200 AU per year, the first crossing to another star could be made in approximately 3,500 years, the life span of the Egyptian Empire. "A more robust propulsion system that enabled a similar trajectory toward higher declination stars such as Alpha Centauri," McNutt notes, "could make the corresponding shorter crossing in a correspondingly shorter time of ~1400 years, the time that some buildings have been maintained, e.g., Hagia Sophia in Constantinople and the Pantheon in Rome. Though far from ideal, the stars would be within our reach."

Why Epsilon Eridani? For one thing, it is a star of high biological interest, surveyed by Frank Drake in his Project Ozma search for extraterrestrial radio signals and studied repeatedly as a possible site for an Earth-like planet. But just as significantly, it lines up well with the plane of the ecliptic. The Earth orbits at thirty kilometers per second, and a spacecraft would have that initial velocity as it begins its journey toward the Sun in the plane of the ecliptic to get a gravity boost to an even higher velocity. Although Epsilon Eridani is considerably further away than Alpha Centauri (10.7 light years as opposed to 4.3), the combination of the initial velocity of Earth and the gravity boost from the Sun gives the spacecraft a significantly higher velocity than either alone.

That makes Epsilon Eridani stand out in comparison to other possible candidates, especially the obvious one. "The trouble with Alpha Centauri," McNutt told me, "is that it's a long way out of the ecliptic. I've got to make a big bend to turn in its direction, so I lose all the angular momentum I gain by choosing a target that's close to the plane of the ecliptic. It was amazing to me when I studied it that it made as much difference as it did."

But no more amazing, surely, than the steep push against the flow of contemporary thought that would be required to think in long time frames. Do we have the sense of purpose as a species to plan and execute missions that would benefit descendants we will never know? The question would haunt my work on interstellar probes, coming up again and again in conversations with researchers. Some argue that a slow probe would inevitably be passed by the faster probes of later generations, but progress is not always linear. Using the same logic, we who have already been to the Moon should have been on Mars by now. There will come a time when we launch something in the teeth of possible futures; we'll also design it to do plenty of good science along the way.

And while the emphasis in modern culture is surely on beating the clock, projects do remain that are multi-generational by design. In Pune, India, scholars composing an authoritative dictionary of Sanskrit have labored for three generations—and six volumes—thus far and have succeeded in finishing only the first letter of the Sanskrit alphabet. Given that the long-dead language, the ancient literary language of Hindus, is laced with puns and infolded meanings that defy even the attempts of scholars to untangle, the completion of this project is surely decades—and probably centuries—off. Yet the work of the Deccan College dictionary project is crucial for understanding aspects of Indian culture, since Sanskrit remained in wide use until A.D. 1100 and was the tongue of scholars and priests.

There is something deeply preservative, and deeply human, that guides such projects. In a similar way near China's Yunju monastery, Buddhist monks began preserving their scriptures in stone in an era when what is now Beijing province was the scene of widespread book burnings. Over 14,000 stone tablets are the result, an effort beginning in the early seventh century A.D. and continuing for a thousand years.

Stewart Brand writes that the idea of a canon, a selection of the best that is known and thought, a selection that is meant to be passed from generation to generation, is one of the great instruments of civilization. In that sense, a multi-generational star probe could be construed as a kind of time capsule, and just as the Voyager spacecraft carried a golden disc filled with the songs and thoughts and images of Earth, so a Centauri-bound craft would carry, if only in its engineering, the legacy of its designers.

But hopes for a faster mission have hardly been abandoned. Ralph McNutt's precursor probe is designed as a testbed for interstellar technologies as well as a science mission that can answer fundamental questions about the nature of the interstellar medium. Along the way, such a probe could examine the disposition of interstellar hydrogen as a way of evaluating the possibilities of a Bussard-style ramjet, the kind of vehicle that could reduce flight times to nearby stars to decades instead of millennia. The goal of making an Alpha Centauri crossing within the lifetime of a researcher remains alive not only in offices searching for major innovations—NASA's Breakthrough Propulsion Physics project (recently terminated but hopeful of future funding) and its Institute for Advanced Concepts are just two of these—but in the scientific literature, where fast-mission concepts abound. Of all of these, none is more famous than Britain's Project Daedalus, the most comprehensive starship design study ever attempted, and still the most entertaining in its origins.

The work of the British Interplanetary Society, Project Daedalus was designed around a fusion drive that uses tiny pellets of fusion fuel instead of nuclear bombs. And unlike the laboratory environment that had spawned so much earlier work on propulsion systems, Daedalus was developed in informal meetings and conversations in English pubs. "Imagine designing a starship while you're sitting around in a bar," said NASA's Geoffrey Landis. "I mean, that's just an incredibly ballsy thing to do!" But not a completely unique thing. In Germany, the Verein für Raumschiffahrt (Society for Space Travel) had been founded in a tavern called the Goldnen Zepter (Golden Scepter) in Breslau. Counting Wernher von Braun among its members, the VfR had a similarly gutsy notion that passion and hard work could lead eventually to manned spaceflight.

Landis is himself a science fiction writer, the author of *Mars Crossing* and a collection of short stories called *Impact Parameter*, as well as one of the rising stars of interstellar theory when he is not working his day job at the Glenn Research Center near Cleveland. His job description covers everything from NASA's wildly successful Mars rovers to photovoltaic systems for generating power in space, and advanced concepts that range from laser-driven solar sails to cosmic strings and wormholes. One suspects a few starship designs may have been jotted down by his prolific pen in pub-like venues at science fiction conventions, as well as the far more academic gatherings like the Space Technology and Applications International Forum held annually in New Mexico, where deep space propulsion ideas flow like wine. So respected is Landis among those who dream of the stars that Robert Forward, upon turning his attention to space tethers at age seventy, specifically tagged him as his successor in the field of interstellar studies.

In an earlier time, a man like this would likely have been one of the British Interplanetary Society's charter members. Like Landis, the BIS has a history of probing the unthinkable, whether in pubs or elsewhere. Founded in 1933 by P. E. Cleator, the Society was first headquartered in Liverpool before moving to London in 1936. It was an early advocate of manned travel to the moon and planets whose members included, in addition to Arthur C. Clarke, such visionaries as fellow science fiction writer John Wyndham and Val Cleaver, an engineer who in the 1950s played a role in a British intermediate-range missile program called Blue Streak. The Society routinely met at London's Mason's Arms pub on Maddox Street, a fact that Clarke would commemorate years later in a collection called *Tales from the White Hart.* These stories capture the atmosphere of the early BIS gatherings as the polymath Harry Purvis spins yarns about improbable scientific advances to a bemused clientele.

But there was more happening at the Mason's Arms than ale drinking and leg pulling. The group was intent on working out the practical problems of space flight in an era when the piston-driven airplane still dominated the sky. That work was temporarily interrupted by World War II, but the BIS remained together in the postwar years to make the case for space exploration, playing a role in keeping German rocket scientists who had been recruited by Britain and America connected as their lines of research began to diverge. Fifty years later, its *Journal of the British Interplanetary Society* remains the fulcrum of serious study. "*JBIS* has been the home of advanced concept thinking," says Landis, "for a very long time now."

Project Daedalus was the work of a group of perhaps a dozen scientists and engineers, BIS members all, led by Alan Bond, who first proposed a working study of an interstellar probe in 1972. While papers were proliferating on specialized interstellar topics, no integrated study of a mission had ever been undertaken. Bond's argument was straightforward: Technology had advanced to the point where a realistic interstellar design could be completed without assuming anything more than we already knew about physics.

The target would be Barnard's Star, at 5.9 light years distance a longer journey than Alpha Centauri, but of great interest because at that time there was intriguing evidence of a planetary system there (evidence which has since been discounted, although no one has disproved the idea of planets around Barnard's Star, either). The flight time was fifty years, the maximum mission length Bond deemed acceptable because it allowed the mission to be flown within the average lifetime of a researcher working on the project. Daedalus would accelerate to 12 percent of the speed of light (roughly 36,000 kilometers per second) and perform a flyby of Barnard's Star, making the encounter time within the system a matter of days, although the study included autonomous probes that would be deployed to nearby planets as Daedalus approached the target. As opposed to the pulsed fusion engines of the main craft, the probes would use nuclear-powered ion propulsion systems.

The Daedalus starship's thrust period was envisioned as four years, during which 250 pellets of deuterium and helium-3 would be detonated every second in the combustion chamber. The two-stage craft, both stages using the same fusion-pulse design, was to be constructed in orbit around Jupiter. It would weigh 54,000 tons, of which fully 50,000 were fuel, with a scientific payload of 450 tons. Described in a 1978 supplement to *JBIS* that remains one of the landmarks of interstellar research, the Project Daedalus study was nowhere more breathtaking than in how it envisioned producing its fuel. Deuterium occurs naturally on Earth (one out of 6,000 hydrogen atoms is deuterium), but helium-3 is rare. Thus the key ingredient—30,000 tons of helium-3—would be mined from the atmosphere of Jupiter.

Today there are other fusion concepts, some of them including the use of antimatter as a catalyst, and a host of propulsion ideas using completely different technologies. The candidate systems—solar sails, beamed lightsails, fusion, antimatter, and interstellar ramjets in myriads of variations—are examined in the chapters that follow, along with thoughts about their current development from the scientists who study them. Wildcard concepts of exceeding interest also make an appearance, including attempts to manipulate inertia and pull energy out of raw vacuum. Propulsion may be nine-tenths of the battle, but we will also consider how to make probes that repair themselves, and the intricacies of navigating between the stars. Science fiction has created many an image of such ships. Today's research sometimes evokes those images, and just as often creates an even bolder imagery of its own.

Any one of the advanced ideas being bandied about in laboratories and research centers could produce a discovery that changes everything, so it is unwise to become too attached to a single means of thrust. I had expected this, and knew that today's scientists would have weighed a probe like Daedalus and found countless ways to improve it. What I hadn't expected was the true wildcard in interstellar study: nanotechnology. For if the biggest issue with sending a probe to another star is the vast expense of pushing payload, then finding a way to shrink the payload down to almost nothing may be a prerequisite. The final chapter of this book will examine interstellar probes the size of sewing needles, sent in their billions on a mission of supreme redundancy, probes that can spawn other probes and build their own communications devices. If this is a bizarre future, it is no more strange than Project Daedalus would have looked to the designers of the DC-3.

Even as a starship embodies our dreams, it also casts light on our history. Lord Collingwood (1748–1810) was Nelson's second in command at the battle of Trafalgar, taking command of the fleet when Nelson was mortally wounded. Collingwood knew that the British oak used in ships built during the Napoleonic wars had been planted during the reign of the Stuarts, some two hundred years earlier, and that it had been designed for future use in the Royal Navy. He would go on to encourage plantings of oaks that would be ready for the Navy a hundred and fifty years after his death.

In an era of quickly fabricated materials, we forget our once deep reliance on the slow processes of nature, and the need to adapt to their rhythms. Interstellar flight demands perspectives as profoundly attuned to its goals as those of Collingwood's foresters, and a commitment to what Tennyson once called "the long result of time." Alpha Centauri, or other dazzling, distant targets like Epsilon Eridani or Tau Ceti, may one day help us recalibrate our vision, while reminding us that no worthwhile goal has ever been easy.

CHAPTER 2

THE TARGET: ALPHA CENTAURI AND OTHER NEARBY STARS

The first fictional journey to Alpha Centauri seems to have been made in Friedrich Wilhelm Mader's 1911 novel *Wunderwelten*, translated into English in 1930 as *Distant Worlds: The Story of a Voyage to the Planets*. Mader's huge spaceship was a sphere 146 feet in diameter that used obscure antigravity techniques to travel to Alpha Centauri at several times the speed of light. Other writers had attempted interstellar journeys, in particular the French authors C. L. Defontenay in *Star ou Psi de Cassiopée: Histoire Merveilleuse de l'un des Mondes de l'Espace* (1854), which also used antigravity in its explorations of the planetary system Psi Cassiopeia, and Camille Flammarion in *Lumen* (1872), a novel featuring a journey to Capella. But Mader's starship *Sannah* was the forerunner of all of today's science fictional journeys to the nearest stellar system.

Alpha Centauri is an obvious target. At 4.3 light years, it is the closest stellar system known. Of the two larger stars in the system, Alpha Centauri A, the brightest, closely resembles our Sun in size, absolute magnitude, and spectral type (which reflects the star's color and hence temperature). But both Centauri A and B are close enough to our Sun in their essentials to make them interesting as interstellar destinations, while the far fainter Proxima Centauri—actually Alpha Centauri C but named "Proxima" because it is the closest known star to our Sun—is a small red dwarf of great interest to astronomers because of its periodic flares. (Recent research has raised doubt about whether Proxima is actually part of the Alpha Centauri system or simply a passing star that is not gravitationally bound to the other two.)

Located in the southern constellation Centaurus, Alpha Centauri is also known as Rigil Kentaurus, meaning "foot of the centaur" (drawn from the Arabic *Al Rijl al Kentaurus*). The system may have been an object of worship to ancient Egyptian civilizations along the Nile. In fact, its first emergence in the morning sky at the autumn equinox has been considered a marker for the orientation of several fourth millennium B.C. temples there. Alpha Centauri is of more recent historical interest because of how its distance was first measured. It was in 1832 that Thomas Henderson, a Scottish lawyer who became director of the Royal Observatory at the Cape of Good Hope in South Africa, subjected the system to distance measurements. Centauri was an obvious choice, since south of 40 degrees north latitude, with its two largest stars appearing as a single star to the unaided eye, it is the third brightest star in the sky after Sirius and Canopus. (At 11th magnitude, Proxima Centauri cannot be seen without a telescope.) Henderson was unaware of the efforts of two other astronomers—Friedrich Georg Wilhelm von Struve and Friedrich Wilhelm Bessel—to make the first measurement of the distance to a star, and while he completed his observations first, Henderson's accomplishment was upstaged by Bessel's announcement of the distance of the star 61 Cygni, some 11.1 light years away.

The work of all three men made use of a method known as stellar parallax. Henderson observed the position of Alpha Centauri, waited six months so that the Earth would be on the opposite side of the Sun in its orbit, and then made the measurement again. The star had appeared to move by three-quarters of a second of arc against the background stars, indicating that it was much closer than they were. Using basic principles of trigonometry and the fact that the Earth's orbit is 300 million kilometers (186 million miles) in diameter, Henderson could calculate its distance: 41 trillion kilometers (25 trillion, 476 million miles). Parallax, the displacement of an object as seen from two points, is a method that works for relatively nearby stars (up to about 200 light years away) but fails with more distant ones that show no apparent motion in the six months between observations. The method could be improved if we had a longer baseline, which is one great value of an interstellar probe—its distance from the Sun would allow much more accurate calculations for stars farther away.

If stars are measured according to their absolute magnitude (actual brightness, as opposed to "apparent" magnitude, which is brightness as seen from Earth) and spectral type, a plot emerges called the Hertzsprung-Russell diagram. Named after the astronomers Ejnar Hertzsprung and Henry Norris Russell, whose independent work led to its discovery, the diagram shows that most stars on such a plot fall into a more or less diagonal line sloping downward called the "main sequence," showing that luminosities drop off as surface temperature decreases. Above the main sequence are large, cool stars, and below are the dim, small white dwarf stars. Where a star is plotted on the Hertzsprung-Russell diagram tells us much about the star's stability. Once the Sun, for example, leaves the main sequence in roughly five billion years, it will enter a red giant phase and, having burned the bulk of its fuel, will end its life as a white dwarf. When choosing a mission destination, we clearly want to focus on main sequence stars similar to the Sun whose stability suggests that any planets around them might harbor life.

Ranging from hottest to coolest, the main sequence spectral types are designated by the letters O, B, A, F, G, K, and M. Essentially this range represents stars at different temperatures and atmospheric pressures, but made of the same chemical components. In terms of color, the O and B stars at the top of the main sequence are the bluest in color, while the type G stars are yellow, the K stars orange, and M stars red. Our Sun is a G2-type star near the middle of the main sequence (each spectral type is further divided from 0 to 9, with 0 being the hottest and 9 the coolest). Seventy percent of the stars in our galaxy are type M, fifteen percent type K, with only three percent type G. The Sun is thus larger and brighter than average; in fact, the bulk of G-type stars are cooler than the Sun.

If we are choosing among the relatively low percentage of stars similar to our Sun, Alpha Centauri A fits the bill in most respects. It is a G2 star whose temperature and color are comparable. Centauri B is not distant on the main sequence, either, being an orange K1 star slightly smaller than the Sun. Only Proxima Centauri is significantly farther away on the main sequence. This M5 star is tiny, a faint, cool object with only a tenth the mass of the Sun. While Proxima in itself might be a less interesting destination, it is the closest of all stars, close enough (13,000 AU) to Centauri A and B to explore on the same mission, making it a useful bonus.

The Centauri system is a southern skies object—it can't be seen further north than central Florida or southern Texas. Recent observations at the European Southern Observatory in Chile have combined the light from two telescopes separated by about a hundred yards to study the Alpha Centauri system. The technique, known as interferometry, creates the effect of a much larger light-gathering surface than either telescope possesses by itself. Called the Very Large Telescope Interferometer, the system has provided the first direct measurements of the size of each Centauri star. Its findings show that Centauri A could not be more Sun-like, although the Centauri system, at 4.85 billion years, seems to be somewhat older than the Sun, which is approximately 4.6 billion years old. Centauri A is calculated to be 1,708,000 kilometers across (1,061,000 miles), making it 1.227 times the diameter of the Sun. Centauri B is 1,204,000 kilometers across (748,100 miles), some 0.865 times the Sun's diameter. The observatory has also measured Proxima Centauri, finding it to be seven times smaller than the Sun.

Given the distances between our Sun and the stars, choosing the nearest one as the target for our first robotic interstellar probe seems an obvious recommendation. But Alpha Centauri has more going for it than simple proximity. The mean separation of the two major stars is 23 AU. That means that if the Sun had a binary companion at the same distance, that star would be slightly farther away than the orbit of Uranus, and both day and nighttime skies would be forever altered as the two stars moved in their orbits. Planets in stable orbits close to each star may be possible. From an Earth-like planet around Centauri A in an orbit similar to ours, the second star would appear at an apparent magnitude ranging from –18.1 to –20.6.

Apparent magnitude is a measure of the brightness of a celestial object to the observer. Almost all objects in the night sky have a positive magnitude, a system in which the higher the number, the fainter the object. The faintest stars visible to the naked eye have an apparent magnitude between 6 and 7, while the faintest stars visible through Earth-bound telescopes have an apparent magnitude around 25. As we move into the negative side of the scale, brightness levels increase. The brightest star visible from Earth, Sirius, is magnitude –1.5. The planet Venus shows an apparent magnitude of –4.4, while that of the full moon is –12.6. The brightest object in our sky, the Sun, has an apparent magnitude of –26.8. Clearly, a Centauri B-type star in an equivalent orbit around our Sun would dominate the night sky, brightening and dimming as the two stars orbited their common center of mass. A planet around Centauri B would see Centauri A even brighter in its night sky, with an apparent magnitude up to –21.9, and its daytime appearances would likely yield double shadows.

So let's imagine ourselves on that Earth-like planet around Centauri A for a moment. In a binary system, both stars move about a common center of mass. And because the orbits of Alpha Centauri A and B are elliptical, the perspective we would have from our planetary setting would show Alpha Centauri B sometimes as far away as Pluto, at other times as close as Saturn. Think of the orbit of Centauri B, as envisioned from the Centauri A system, as being intermediate between a more circular planetary orbit and a more elliptical cometary orbit. At its farthest point, Centauri B would shine as a tiny orange disk with 100 times the brilliance of a full moon, far and away the brightest star in the sky. As Centauri B moved in its orbit to close the distance between the two stars, it would reach its closest point in forty years. Now the star would shine with the light of 1,400 full moons, although still 300 times fainter than Alpha Centauri A.

A planet around Centauri B would see even more spectacular visual effects. From its perspective, it is now Centauri A that moves in a long elliptical orbit around Centauri B. But Centauri A is more than three times as bright as Centauri B. So at its closest point, Centauri A would shine 5,000 times brighter than our full moon, fully 1/30 as bright as Centauri B.

As for Proxima Centauri, it would be a dim star of magnitude 3.7. Dim enough that we would be unlikely to pay any attention to it. Isaac Asimov once remarked that Proxima would only get serious attention from Centauri astronomers when they realized that it possessed a large parallax, a fact that would tell them that it was either a member of their own stellar system, or an independent star happening to pass close by. Recent studies of Proxima Centauri have tended to favor the latter hypothesis, as its radial velocity seems too great for Proxima to remain within the Alpha Centauri system. Moreover, its frequent flare activity suggests that Proxima may be younger and thus may have emerged from a different stellar nursery than Centauri A and B.

But can planets form around double or triple star systems? And if they can, would the orbits they achieve be stable? One problem is that the system's orbital eccentricity may disturb planetary orbits. Orbits are rarely perfect circles, but usually ellipses that are more or less elongated. In astronomical terms, a perfectly circular orbit has an orbital eccentricity of 0, while an extremely elongated orbit is close to a value of 1. Because Centauri A and B display an orbital eccentricity of 0.52 around their common center of mass, their separation changes during their 80-year orbit around it, ranging from 11 to 35 AU. These orbits must be thoroughly analyzed to make sure they are not disruptive to planets. One recent study is encouraging, suggesting that the formation of terrestrial-size planets cannot be ruled out at least around Alpha Centauri A. Given that current theories of planetary formation are being challenged by the exciting findings of modern-day planet hunters, and because such theories have largely been based on single-star systems like ours, we are still working on the definitive answer to whether planets can form.

If planets can form in double- and triple-star systems, would their orbits be stable? In a 1996 paper first presented at the Practical Robotic Interstellar Flight: Are We Ready? conference two years earlier, Alan Hale described two scenarios for stable planetary orbits within binary star systems. In one, the two stars are relatively close, allowing planets to orbit both of them simultaneously. In the other, the two stars are widely separated and the planets orbit one of the two. Each of the stars, in other words, might have its own planetary system. Examining binary star systems that might contain a planet like Earth in a stable orbit, Hale noted that for a planet to orbit both stars at once (assuming a circular planetary orbit), their separation must be less than 0.1 AU, or 9.3 million miles. If the planet were to orbit one of two widely separated stars, those stars must be at least 10 AU distant from each other. Twenty-two binary star systems met the latter criteria in Hale's study of nearby stars, with Alpha Centauri topping the list. So while we do not yet know if planets exist around Alpha Centauri, we have reason to believe that planetary orbits could be stable there.

The assumption of stability was further refined by Paul Wiegert and Matt Holman in 1997. The two astronomers found that stable orbits can exist within 3 AU of either Centauri A or B. They went on to calculate the habitable zone around Centauri A, finding it to be from 1.2 to 1.3 AU, while that around Centauri B is 0.73 to 0.74 AU. A second stable zone was found to occur beyond 70 AU, encompassing both stars. These questions will not be resolved unless we are able to obtain images of a Centauri planetary system, but given the results of Wiegert and Holman, it is possible that both Centauri A and B might contain rocky inner planets like those around our Sun out to the orbital distance of Mars (the orbits of outer gas giants like those in our solar system would clearly be disrupted). Within this range, temperature zones where liquid water is possible could be found on planets orbiting either star.

A later European study was even more emphatic: "These terrestrial-size planets form within the region of planetary stability found by Wiegert and Holman (1997) and are possibly stable over the age of the system," wrote astronomers Hans Scholl and Francesco Marzari. "Some of them lay within the habitable zone of the star and, possibly, could harbor life."

The obvious inference—and it is a logical one—is that to be the target of an interstellar probe, a star must reveal the possibility not just of planets, but also of life-bearing planets. This would change if interstellar probes were cheap and numerous, in which case missions to investigate a wide variety of stellar types would make sense. But given that the problems in building interstellar probes are so immense and the costs so potentially high, we are unlikely to launch one unless we have reason to think that life-bearing planets orbiting our target star are possible.

Add to Alpha Centauri's possible biological interest the fact that Proxima Centauri represents a scientific target in its own right. If it is indeed gravitationally bound to Centauri A and B, Proxima's orbital period around the duo is probably around half a million years. It is 19,000 times fainter than the Sun (faint enough to have eluded detection until 1915) and is a flare star, doubling its light output in a matter of minutes in ways that fascinate astrophysicists. The detection of planets around such a star would be unusually interesting, but even in their absence, Proxima is a desirable destination.

For it turns out that seven out of ten stars in our galaxy are red dwarfs like Proxima Centauri. Finding that planets could form around them would thus significantly affect our view of how many planets exist. It is interesting to speculate on what a hypothetical planet around Proxima Centauri might be like. One study performed with readings from the Hubble Space Telescope collected over forty sets of data on the star field around Proxima. The study found a 77-day change in the proper motion of Proxima that could conceivably be caused by the gravitational tug of a planet near the size of Jupiter, orbiting less than half the distance of Mercury from our Sun. A later study, also using data from Hubble, may have observed a planetary companion (perhaps a so-called "brown dwarf," intermediate between a star and a planet) roughly half the distance between Earth and the Sun. Later observations at the European Southern Observatory leave the question unsettled, finding no evidence for planets larger than about 80 percent of the mass of Jupiter in nearby orbits. In short, the accuracy of our measurements does not yet allow us to rule out the existence of planets around Proxima Centauri, but we are beginning to be able to rule out planets of at least a certain size.

In order to be warm enough to have liquid water, a planet around Proxima would have to orbit quite close to the star, a fraction of the distance between Mercury and the Sun. Life would be possible, even though Proxima is a flare star. Imagine a planet with a vertical axis of rotation with respect to its orbit and whose rotation has the same period as its orbit. This would keep one side in perpetual daylight, the other in darkness. Its nearness would make Proxima's frequent x-ray-laden flares dangerous but not necessarily fatal to life, and the right atmosphere could theoretically make the planet habitable. Simulations of the atmosphere of a hypothetical red dwarf's planet conducted in the 1990s at NASA's Ames Research Center showed that even a relatively thin atmosphere could circulate heat to the dark side of the planet, keeping it warm enough to prevent its gases from freezing out. Add oceans and you have the prospect of a liquid water layer beneath ice on the planet's dark side, heated by geothermal activity within the planet itself. On the day side, Proxima would be a dim ember with most of its energy being emitted in the infrared. Even so, some form of photosynthesis might evolve there.

What a strange world this would be. On the day side, the sun would be a stationary red ball. There being no axial tilt, there would be no seasons. Temperatures at the equator would be the highest, dropping off as we moved in any direction. The coldest point would be at the center of the night side. If life did evolve on such a world, finding a way to adapt to an environment in which infrared—not visible light—predominates, it is possible that the star's periodic flare activity could actually serve as an evolutionary stimulus. Intriguingly, red dwarf stars have long life spans, perhaps one hundred times the ten billion or so years allotted our Sun. That would give evolution plenty of time to work its magic. Do such planets exist? Suggestive results from a binary system called CM Draconis, composed of two red dwarfs, may indicate the presence of Earth-size worlds there, but scientists working on the project say they need more data before they can reach even a tentative conclusion.

We have only one solar system we can study closely, and ours is a place of enormous contrasts, from the outer gas giants to the inner rocky worlds, with a zone of asteroid debris between the two categories of planets, and an apparently huge number of icy objects in the Kuiper Belt beyond the orbit of Pluto. It will be some time before we know definitively whether life can form in environments like oceans trapped under ice (Europa) or deep under the surface of hostile deserts (Mars), and we cannot absolutely rule out even stranger life forms in the planetary atmospheres of gas giants themselves, not to mention conceivable biospheres within the Kuiper Belt.

But it makes the most sense to look for life under conditions like our own. On that score, it was Stephen Dole who called our attention to the "habitable" zone in which, around distant stars, life might be found. Writing for a RAND Corporation study in 1964 (and later in a popular account with Isaac Asimov), Dole summarized the requirements for an Earth-like world in terms of its mass (greater than 0.4 times and less than 2.35 times the mass of the Earth), period of rotation (less than about 96 hours to prevent temperature extremes), and age (greater than three billion years to allow for the appearance of complex life forms). To allow a zone in which such a planet could exist, Dole developed the concept of a stellar ecosphere, "... a region in space, in the vicinity of a star, in which suitable planets can have surface conditions compatible with the origin, evolution to complex forms, and continuous existence of land life; and, in particular, surface conditions suitable for human beings, together with the whole system of life forms on which they depend."

Dole's work was all about ideal conditions rather than marginal ones. Thus a promising planet should exist within a temperature zone that allows for the existence of liquid water. Our Earth obviously fits that bill, and Dole calculated that it could be slightly closer to the Sun and still do so, with an outer range perhaps as far as 1.2 AU from the Sun, producing a habitable zone .30 AU in depth. But that is for stars much like the Sun. Other spectral types would have different habitable zones depending on their size and temperature. Alpha Centauri B, for example, is a K1-type star, an object whose habitable zone might extend as close as .60 AU because it is dimmer and cooler than the Sun. And as we have just seen, the M-class Proxima Centauri is so cool that producing Earth-like temperatures anywhere nearby would involve orbiting much closer than the orbit of Mercury.

The concept of a habitable zone is widely accepted, and Dole, who believed planets could exist in stable orbits around both Alpha Centauri A and B, liked the odds, calling the Centauri system "... the highest probability for any star or star system on our list ..." What is controversial is the size of the habitable zone around a given stellar type—if the zone is tiny, the likelihood of planets falling exactly within it decreases, with obvious implications for the existence of life on other worlds. Recent work has also questioned whether there is a "galactic habitable zone," not too close to the Galactic core to be pounded by intense radiation, and not so far toward the edge to be metal-poor.

We have had to learn not to be too doctrinaire about planetary orbits and the conditions on planets unlike those of our own solar system. Before he consulted with Apollo 11 astronaut Buzz Aldrin and science fiction author John Barnes about a possible Centauri system for their novel *Encounter with Tiber*, interstellar theorist Gregory Matloff had assumed that no Jovian planets could exist within 1 AU from a star like Alpha Centauri A or B, as the novel demanded. But running the numbers in the early 1990s, Matloff realized that hydrogen/helium atmospheres could indeed exist for billions of years at such distances. Remarkably, that finding was confirmed within the decade as planet hunters discovered numerous so-called "hot Jupiters" orbiting their stars in orbits tighter than this. The G8-class binary system 55 Cancri in Cancer is a case in point. A planet detected around 55 Cancri A, presumably a gas giant, is 0.84 times as massive as Jupiter, yet orbits the star in a mere fourteen days at 28 percent the distance of Mercury from the Sun.

Whether or not the Alpha Centauri system has planets is a question of considerable moment not just for the launching of a possible robotic probe, but for the wider issue of life in the universe. Spectroscopic studies of nearby solar-type stars show that about two-thirds of them have stellar companions, and some estimates indicate that up to 80 percent of all stars in the sky are members of binary systems. Finding planets within binary systems and around red dwarfs is therefore a crucial factor in how widespread terrestrial-like planetary systems may be. Early evidence on planetary formation within binaries is scant, but one interesting planet has been found around the binary system Gamma Cephei.

It was in October of 2002 that astronomers at the McDonald Observatory at the University of Texas announced the find. Presenting their results to the American Astronomical Society's Division of Planetary Sciences meeting in Birmingham, Alabama, the Texas team said the planet they had discovered orbited the larger star of the Gamma Cephei binary system, about forty-five light years from Earth in the constellation Cepheus. Not quite twice as massive as Jupiter, the planet orbits a star 1.6 times as massive as the Sun at a distance of 2 AU. Interestingly enough, the second, smaller star of this system is about 25 AU away from the larger star, comparable to the distance of Uranus from the Sun, and not far off the mean distance between Alpha Centauri A and B. While the Gamma Cephei planet is not the first found in a binary system, it is the first discovered around a close binary pair—that is, the other planets were in systems whose stars were so widely separated that each would have little effect on the possible planetary system of the other.

Planet hunting is filled with "firsts" as we begin to use advanced astrometric techniques to measure the wobble in a star's path through space as it is gravitationally affected by planets. The wobble is tiny but discernible with the exquisitely sensitive equipment now being used to detect the Doppler shift produced in the star's light by its motion. In such ways, we can already confirm the existence of well over one hundred planets around various stars. Other methods, like directly imaging the passage of a planet across the face of a star whose planetary system lies edge-on to us, are still in their infancy and likely to yield their own trove of new planets.

Geoff Marcy, a professor at the University of California at Berkeley and principle investigator for NASA's Space Interferometry Mission, has co-led (with Paul Butler) a team that has chalked up more than seventy planets outside our solar system. Marcy remembers what he calls the terror and excitement of his first planetary discovery. "I'll never forget the morning of December 30, 1995. I was at home, preparing for New Year's Eve, and my collaborator, Paul Butler, called me up and said just three words—'Geoff, come here.' He was already at the office. It was a Saturday morning. I drove immediately to Berkeley, and there on the computer screen was a plot showing the wobble of [the star] 70 Virginis. We had been looking for planets around stars for eleven years without a single success, and there on the computer screen was the first planet we had ever discovered. It was a fantastic moment."

A far greater moment lies ahead, perhaps in the next ten years. For Marcy believes that we will know whether other Earth-like planets exist around some 200 nearby stars within the next decade. That probably wouldn't surprise former NASA administrator Dan Goldin, who pronounced finding such worlds to be a goal in 1999 when he spoke to the American Astronomical Society. In the same presentation in which he had sketched out a hypothetical interstellar probe, Goldin imagined how future space telescopes might bring images of distant solar systems to the classrooms of the mid-twenty-first century: "When you look on the walls, you see a dozen maps detailing the features of Earth-like planets orbiting neighboring stars," Goldin said. "Schoolchildren can study the geography, oceans, and continents of other planets and imagine their exotic environments, just as we studied the Earth and wondered about exotic sounding places like Banghok and Istanbul . . . or, in my case growing up in the Bronx, exotic far-away places like Brooklyn."

Images like these would provide fascinating confirmation of terrestrial environments around other stars, and would doubtless tell us whether or not such planets contained life. But apart from their raw scientific interest, pictures of distant planets would also supply what we need to know to choose a target for our first interstellar probe. None of the planets so far found are terrestrial—they are all gas giants, in most cases far larger than Jupiter, the behemoth of our solar system, and they tend to occupy elliptical orbits close to the stars they circle. The limitation is simply that of our measuring techniques, which are not yet refined enough to locate small, rocky worlds like Earth. With the relatively gross methods of our current search, we have no definitive information about planets around Alpha Centauri, but a Jupiter-like planet has been found circling nearby Epsilon Eridani (with some evidence of a possible second planet), some 10.5 light years from Earth. We will not launch an interstellar probe until we have surveyed all nearby stars for both kinds of worlds.

Yet even with today's methods, significant progress is being made. Astronomers using the 3.9-meter Anglo-Australian Telescope in Siding Springs, Australia, announced in the summer of

2003 that they had discovered a planet much like Jupiter orbiting a star similar to our Sun. Speaking in Paris to the Extrasolar Planets: Today and Tomorrow conference, team leader Hugh Jones described the new planet as twice as massive as Jupiter, circling the star HD70642 about every six years. About ninety light years from Earth, this star appears to terrestrial observers as a 7th magnitude object in the southern constellation Puppis. If the new planet were in our solar system, it would occupy a position between Mars and Jupiter. Significantly, and unlike many other early planetary discoveries of gas giant worlds close to their parent star, this system has no other giant worlds orbiting between it and HD70642. The similarity to our own solar system suggests the possibility of small, rocky worlds in the habitable zone inside the gas giant's orbit.

Moreover, and also unlike many other planetary discoveries, the planet around HD70642 moves in a nearly circular orbit. We have thus found for the first time a solar system that may be much like our own. The only one comparable is the system around 47 Ursa Majoris, first detected by Paul Butler and Geoff Marcy in 1996, and later extended by them with the discovery of a second planet. That system includes one gas giant in an outer orbit and another at a distance of approximately 2 AU (both orbits would fit inside the orbit of Jupiter in our solar system).

The discovery of an Earth-like world around a nearby star would focus intense attention on building a probe to study it, particularly to discover whether life exists there. But in the case of Alpha Centauri, the complex orbital mechanics of the two stars masks the far more subtle effects of any planets, meaning we must turn to methods other than Doppler-detected wobbles. The installation of a new device called the Advanced Camera for Surveys (ACS) aboard the Hubble Space Telescope may at least answer the question of whether Alpha Centauri has planets, though it will be unable to detect worlds as small as Earth. Installed in Hubble by the Space Shuttle Columbia in March of 2002, the ACS boosts Hubble's efficiency by a factor of ten, using three electronic cameras and a variety of filters and detectors. Spectacular proof of its prowess is an image called the Ultra Deep Field, a breathtaking vista of over 10,000 galaxies, some of them more than 13 billion light years distant, that was unveiled to the public in early 2004.

An early project of the ACS science team at Johns Hopkins University and the Space Telescope Science Institute in Baltimore will be to study the Alpha Centauri system using an Aberrated Beam Coronagraph. This instrument uses two occulting masks to block the otherwise overwhelming light of the nearby star as a planetary search is mounted. The investigation draws on Wiegert and Holman's work on stable orbits in the Centauri system, and proceeds under the assumption that if a Jupiter-size planet exists around one of the stars of Alpha Centauri, it should be possible to obtain its image.

Finding planets visually pushes even Hubble's latest upgrades to their limit, and in any case, Hubble's own future is now clouded by funding as well as safety concerns that may cause the telescope to be abandoned. But other possibilities are under study, including plans for a mission called Terrestrial Planet Finder, whose final design will be chosen in 2006 (the actual mission will not fly until at least six years later). Among the options for TPF: an eight-meter visible light telescope and new imaging technologies, including an infrared coronograph, a 28-meter-wide mirror configured to block out the light of the star while detecting the light of its planets. Also being considered is an infrared interferometer that combines the light from five separate spacecraft, each containing four eight-meter telescopes, along with a separate eight-meter telescope that relays the collected light to a "combiner" spacecraft. The resolution obtained would presumably be spectacular. In fact, it would be similar to what could be achieved with a telescope as big as the parabola defined by the spacecraft themselves; imagine a mirror a thousand kilometers (621 miles) across.

The TPF mission should be able to measure planets as small as the Earth in the habitable zones around other solar systems. Using spectroscopic analysis, chemists and biologists will then be able to look for "biomarkers," measuring the relative amounts of gases like water vapor, carbon dioxide, ozone, and methane to find out whether such planets contain life, or did so in the past. TPF will also be able to study the development of solar systems from the disks of gas and dust surrounding recently formed stars. As currently envisioned, the instrument will be capable of sending information about possible terrestrial planets around 150 stars within 45 light years of Earth. An even larger mission called Life Finder has been proposed as a follow-up; it would extend TPF's reach to thousands of stars and sharpen the accuracy of our observations of planetary atmospheres by examining their spectral signatures.

But astronomer Webster Cash of the University of Colorado at Boulder thinks the Terrestrial Planet Finder can be enhanced far beyond what infrared coronographs and even huge, space-based conventional optics can yield. Not only that, his planet-imaging system would significantly reduce the engineering challenges created by the existing TPF design. Terrestrial Planet Finder's 28-meter mirror would have to be flat to a tolerance of a single atom. Nor could there be the slightest variation in its reflectivity, which must be uniform to a tolerance of 99.999 percent. Such precision is demanded because of the enormity of the problem: If we were trying to snap a photograph of the Earth from Alpha Centauri, our planet would appear ten to twenty billion times fainter than the Sun, depending on the Earth's phase. Terrestrial Planet Finder is built around the premise of extracting a useful image out of that glare, but can its extraordinary mirrors actually be built?

Cash's credentials at imaging distant objects were already impressive before he began advocating direct imaging of Earth-like planets around nearby stars. The x-ray telescope he and his team developed provides a million-fold increase in x-ray resolution over earlier systems, potentially letting us see matter spiraling into black holes. Being considered for a mission called MAXIM (Micro Arc-Second X-Ray Imaging Mission), the telescope might ultimately involve thirty or more spacecraft flying as a colossal interferometer that combines the image from each. Such a telescope would be capable of imaging an object the size of an automobile at the center of the Milky Way.

Having studied the Terrestrial Planet Finder concept, and drawing on his MAXIM work, Cash now advocates a basic planet-finder that uses two spacecraft. The first is nothing more than a sheet of black plastic—as much as a kilometer across, and shaped like an umbrella—with a hole in the middle. The hole is an aperture—an opening in a lens that admits light. The system effectively creates an enormous pinhole camera. It may be hard to imagine such a homely object probing distant stars; after all, in terrestrial terms, a pinhole camera is nothing more than a light-proof box with a black interior and a small hole in the center of one end and film at the other end. Keeping a pinhole camera absolutely still while exposing the film results in surprisingly good images. The working premise is sound: Light coming through the pinhole is inverted and reversed, producing a faint image that requires long exposure times to be captured on film.

In such a camera, it is the pinhole that is the lens, whereas a standard camera uses a conventional optical lens because it needs the much larger aperture for faster exposures. But given the issues involved in creating enormous mirrors to the TPF tolerance, Cash sees this simple technology as adaptable for space missions. The pinhole on his space-borne plastic sheet would be meters across. His second spacecraft would be placed tens of thousands of kilometers away. It would consist of a standard telescope with an aperture the same size as the pinhole in the sheet. From that distance, a pinhole image of a distant planetary system would be spread across the sky. Looking at an Earth-like planet from a distance of ten parsecs (about 32 light years), an astronomer would see an image of the planet fully ten meters away from its star on the focal plane of his pinhole device. The dark sheet would suppress the background light. Moving around in the focal plane would allow the telescope to see the target planet, or other planets in the system, or the central star.

Forget the tight tolerances of the TPF mirrors. "Notice the requirements on Terrestrial Planet Finder," Cash told me. "It demands perfect reflectivity and no variation in phase from the mirror. A pinhole camera does that automatically because the light just goes through the hole. So the pinhole camera has perfect phase control. And then you just shape it to suppress the diffracted light from the central star. I'm working with some scientists at Princeton who know how to do that." Cash believes his two-spacecraft system will be able to provide spectroscopic analyses of Earth-like worlds that will help us understand their atmospheres, while later generations, with perhaps fifty sets of spacecraft using interferometry to pool their data, will be able to provide breathtaking visual close-ups. "There is no necessary limitation on this optical system," he added. "We might create a view as close as a hundred kilometers, looking at weather systems, oceans, continents." The only issue would be that due to its orbital inclination as the planet rotates around its star, we wouldn't be able to map every inch of such a planet, but only the surface it presented to us from our viewpoint. And because the spacecraft would have to hold their position along the line of sight to the planet, they would have to be placed at the L2 point, one of five orbital positions first studied by the eighteenth century mathematician Joseph Louis Lagrange where Sun, Earth and spacecraft stay in place relative to each other as they move through space.

These seem like minor inconveniences given the magnitude of the knowledge we would gain. Cash's concept is being considered by NASA right now, and may have an impact on the final design of Terrestrial Planet Finder, perhaps allowing for a more efficient and probably far less expensive mission. A certain irony of the situation is not lost on Cash. "When I heard Dan Goldin say that about visualizing Earth-like planets a dozen or so years ago, I just thought he was nuts." Cash laughs. "And now I'm proposing to do it. I think we can do it inside twenty years if we can get NASA to commit to this. Because there is no piece of technology they don't have right now."

But whatever its final form, Terrestrial Planet Finder will build on two earlier missions, for the drive to image Earth-like planets is gaining momentum. One mission, called Kepler, is designed to detect extra-solar planets by looking for systems whose planets pass in front of their parent stars. This "transit method" should see light from the star dim as a planet passes in front of it, allowing scientists to measure both the size of the planet and its orbit. Scheduled for a 2007 launch, Kepler will use a 1-meter diameter telescope called a photometer to measure the exceedingly subtle changes in brightness produced by these transits. Given that most stellar systems will not be aligned properly to allow the viewing of a transit, the spacecraft will monitor thousands of stars, looking for the characteristic signature of a planetary crossing. The Space Interferometry Mission, scheduled for a 2009 launch, will use optical interferometry to combine the light from multiple telescopes.

Both missions will eventually be supplemented by the James E. Webb Space Telescope, to be launched in 2011. Considered the successor to the Hubble Space Telescope, the JWST will use a 20-foot primary mirror protected by a sunshield to make infrared observations of planetary atmospheres and perhaps biological activity on extra-solar planets.

For its part, the European Space Agency is studying two missions. The first, called Eddington, is a precision photometer that measures tiny changes in the brightness of stars, some of which may be caused by the transit of planets across the stellar disk. Capable of detecting planets half the size of Earth, Eddington was scheduled for launch in 2008 but may not fly due to budget troubles. A later mission, called Darwin, is conceived as a cluster of spacecraft working with 2-meter telescopes to examine infrared light.

Infrared is helpful because in visible light a star would completely mask the billion-times weaker light from an Earth-like world. In the infrared region of the spectrum (with wavelengths from roughly one micrometer to one millimeter), the difference is "only" a million, meaning it will be easier for the image to be extracted from the background light. Infrared observation has already had some successes, particularly in the Infrared Astronomical Satellite (IRAS), a collaborative project between the United States, United Kingdom, and the Netherlands that was launched in 1983. IRAS made observations of clouds of debris around several stars, suggesting planetary systems (at least in their early stage of development) around Vega, Fomalhaut (Alpha Pisces), Beta Pictoris, and Epsilon Eridani. Darwin will deploy six spacecraft in a 100-meter formation, combining their observations to achieve the imaging power of a much larger mirror, and feeding these to a processing spacecraft and then a communications relay to return the images to Earth. The key to Darwin is formation flying, which will be tested in a precursor mission to be launched in 2006. Darwin itself is not expected to fly before 2014.

Such missions will tell us much about what is around Alpha Centauri. If it proves to have no planets, we will doubtless set our sights on other nearby stars that do. There will be no shortage of targets. In the spiral arm of the galaxy in which we reside, we can find 103 known stars—in a variety of spectral types—in 78 star systems within a range of 21 light years. Among Sun-like stars, there are approximately 100 in the G1 to G3 class within a sphere with a radius of 70 light years. As the Sun is a G2 star, this implies 100 Sun-like targets around which Earth-like planets may have formed. All will presumably be the subject of telescopic observations from spacecraft like Darwin and Terrestrial Planet Finder.

One intriguing possibility is that our Sun is itself a member of a binary star system, with an as yet undiscovered dwarf star orbiting Proxima-like at such a distance that we might discover it only through its effects on nearby space. Such a star, called Nemesis, has been hypothesized because of the regularity of species extinction events on the Earth, the theory being that on its closest pass to the Sun every twenty-six to thirty-four million years, Nemesis would create instabilities in the Oort Cloud that would cause a shower of comets to cascade toward the inner solar system, some hitting the Earth to create acid rain, ozone loss, and global winter. It is possible to work out an orbit that would account for these mass extinctions, one in which Nemesis ranges from 2.8 light years from Earth to as close as 0.01 light years. But the star—or brown dwarf, perhaps, a "failed star"—would be hard to detect because of the dimness of Nemesis and the number of candidate objects whose parallax would need to be examined. In any case, if Nemesis exists, it is likely at the far side (aphelion) of its orbit, since the last mass extinction was thirteen million years ago.

Any number of other stars may exist closer than Alpha Centauri, presumably in the M-class red dwarf category, and thus hard to detect. It was not until 2003 that the red dwarf SO25300.5+165258 was discovered, a star which obviously needs a catchier name to rival Alpha Centauri or Barnard's Star. But this tiny star is only 7.8 light years away, making it the third closest to the Sun. As automated sky surveys continue to update our knowledge of the stellar neighborhood, we can look at more established candidates as possible destinations.

At six light years, Barnard's Star was the target chosen by the Project Daedalus team for their fusion-powered starship. An M5-class red dwarf one-sixth the size of our Sun, Barnard's Star was discovered in 1916 by Edward Emerson Barnard, working at California's Lick Observatory. The star's fast motion against the background stars was obvious when Barnard compared photographic plates taken at different times. Indeed, the star moves so quickly through the constellation of Ophiuchus that it needs only 350 years to cover one degree of sky (put another way, the star moves the width of a full moon in 180 years). If we wait long enough,

Barnard's Star will make an easier target, as it is approaching the Sun at a speed of 87 miles per second, a rate that will take it to within four light years of the solar system in 8,000 years.

Barnard's Star is the closest that can be studied from the Northern Hemisphere, but only through telescopes, as at an apparent visual magnitude of 9.56, it is far too faint to be seen with the naked eye. If we were looking for a truly average stellar target, this one would fit the bill; it is by far the most common type of star in the Galaxy. Of the fifty-two stars within 16 light years of Earth, thirty-five are M-class stars. Only three are G-type, like the Sun and Alpha Centauri A, while seven are K stars like Alpha Centauri B. In that same 16-light year sphere, nine double star systems exist, and two triple star systems.

If we were to reduce interstellar mission candidates to single G- and K-type stars, the leading possibilities would be Epsilon Eridani (a K star with at least one discovered planet 10.7 light years from Earth), Tau Ceti (a G star 11.9 light years away), Epsilon Indi (a K star at 11.2 light years) and Groombridge 1618 (a K star 15 light years away). Epsilon Eridani is particularly interesting. Somewhat younger than the Sun, it shows a dust ring extending between 35 and 75 AU from the star, with a clear zone closer in that may have been produced by planets whose gravity has swept up the dust, in a pattern not dissimilar from what we believe happened in our own solar system. But Epsilon Eridani is a long way out. If we decide that Alpha Centauri is unsuitable and that M-class stars are of less biological interest, then we have more than doubled the range of a mission to the next target. All things considered, life will be much simpler if our searches do reveal planets around Alpha Centauri.

Whether bound for Alpha Centauri, Epsilon Eridani or Tau Ceti, though, an interstellar probe will have to cope with the nature of the so-called interstellar medium. In his novel *Flying to Valhalla*, Charles Pellegrino depicts twin antimatter-driven starships on a mission to Alpha Centauri, one of which is destroyed by a collision with space debris. "Flying through space at significant fractions of lightspeed is like looking down the barrel of a super particle collider," writes Pellegrino in an afterword to the book. "Even an isolated proton has a sting, and grains of sand begin to look like torpedoes." Such problems are one reason why precursor missions of the sort the Jet Propulsion Laboratory has studied make sense. In addition to measuring cosmic and gamma rays, observing the background radiation of the universe, and studying the magnetic fields, temperature, and density of interstellar gas, such missions would help us understand the makeup of dust clouds within and beyond the solar system.

Nearby interstellar space is, after all, a place we know little about. Our knowledge of it comes from observing its effects on starlight and radio waves, and from the light and radio waves emitted by interstellar gas clouds themselves. But we have learned something about how the solar system interacts with it. The so-called solar wind is a stream of charged particles and magnetic fields moving at approximately 500 kilometers per second away from the Sun. As it sweeps outward through and beyond the orbit of the nine known planets, it carves out the heliosphere, a bullet-shaped bubble of hot plasma in which the solar system is enclosed. The surface where the solar wind drops below the speed of sound (roughly 30 to 50 kilometers per second) is called the "termination shock," an area that may be as close as 75 AU. As the solar wind continues to slow, it is turned in the direction of the flow of interstellar gases to form a tail behind the Sun not unlike that of a comet.

The exact shape and size of the heliosphere remain unknown, although data from the two Voyager spacecraft should provide our first indications within the next ten years. Indeed, some scientists argue that Voyager 1 has already reached the termination shock, a view that is still controversial. At the edge of the heliosphere is the heliopause, the boundary between the solar wind and the interstellar medium through which the Sun moves in its rotation around the galactic core. The heliopause is estimated to be anywhere from 100 to 160 AU from the Sun. The motion of the heliopause through the galactic medium produces what is known as the bow shock. The metaphor is apt; think of a boat moving through a harbor, pushing a curl of water out in front as it proceeds. This interstellar bow shock may extend out as far as 230 AU. What the heliopause ultimately defines is the edge of the region where the Sun's influence effectively ends. While it is true that neutral interstellar atoms can penetrate the heliosphere, most of the material within the bubble comes from the Sun. As distant Pluto averages a mere 39 AU from the Sun, it is clear that the actual "boundary" of the solar system may be up to six times further out than the farthest known planet. The discovery of trans-Plutonian planetoids like Sedna, whose comet-like orbit reaches beyond the heliopause at the furthest point of its orbit, further muddles our notion of how far the solar system extends.

The NASA interstellar mission conceived by JPL's Interstellar Probe Science and Technology Definition Team was designed to explore the interaction of the heliosphere with the interstellar medium and to examine that medium itself to understand the astrophysical processes occurring within it. What such a probe finds would have considerable bearing on future missions to Alpha Centauri because the faster a probe moves, the greater the danger from anything it encounters along its route of flight. Even Voyager 2, moving at comparatively slow interplanetary speeds as it approached Uranus and later Neptune, felt the effects of space-borne dust, measuring its impact upon the spacecraft's skin with its plasma wave instrument. This instrument was designed to measure the charged particles inside the magnetic field of gas giant planets, but it can also register a hit when the spacecraft encounters dust, sensing the plasma created by the vaporized particle.

Some scientists agree with Pellegrino that at interstellar speeds a speck of dust might as well be a mountain. When physicist Steven D. Howe, the proponent of an audacious antimatter sail concept (described in Chapter 4) discussed interstellar flight at a panel at the World Science Fiction Convention in Los Angeles in 1996, he went straight to this point. As he told me in a later interview, "I proposed that humans (or anyone else for that matter) will never go to another star through normal space because at the minimum speed of .1c a grain of sand will destroy a ship. If other solar systems have an Oort Cloud and a Kuiper Belt, any entering ship will be blasted."

Not everyone agrees that the dust problem is insurmountable, and a variety of shielding systems have been concocted to address it. But all agree that we need a better set of data for measuring interstellar dust. The Voyager readings were taken with an instrument not designed for dust detection, and we have not yet had the opportunity to corroborate them with other missions. However, the Cassini Saturn mission launched in 1997 carried the Cosmic Dust Analyzer, an instrument that measured interplanetary dust grains beginning with Cassini's gravity assist by the planet Venus in 1999 and lasting until its arrival at Saturn in the summer of 2004. The Stardust mission, which in 2004 collected dust samples from Comet Wild 2 in addition to dust along its route of flight, used an instrument called the Cometary and Interstellar Dust Analyzer.

What we know about the nearby interstellar medium is that its density is far lower than even the best vacuum we can create on Earth, apparently averaging .01 atom of hydrogen for every cubic centimeter of space (the overall average Galaxy-wide appears to be higher, in the range of one atom per cubic centimeter—our solar system seems to be in a pocket of unusually sparse material). Up to 99 percent of the interstellar medium is gaseous, mostly hydrogen with about 25 percent helium and some heavier elements. Dust is rarer still, with one dust particle for every trillion atoms. This mix of gas, dust, magnetic fields, and charged particles is generally found in the form of cold clouds or, near stars, as hot, ionized hydrogen gas. The spectacular visual effects of ionized hydrogen gas (energized by radiation from newly formed stars) appear in the glow of famous celestial objects like the Orion Nebula.

The dust particles are another story. Mostly carbon, ice, iron compounds, and silicates, these particles are normally less than a millionth of a meter across—less than 1/25,000 of an inch, which is about the wavelength of blue light. Because of this, red light passes through dust and gas clouds more readily than blue, so some objects appear redder than we would expect. Their effect is measurable from Earth through an effect called extinction, as light is absorbed and scattered through clouds of dust particles. You can see the same effect at sunset or sunrise, when dust and gas molecules absorb the bluer colors of sunlight to yield redder hues. Our Sun, situated in the Orion arm of the Milky Way, currently lies several thousand light years outside a denser cloud of dust and gas where star formation is more intense. Imagine the solar system passing through an intense dust cloud and you'll see that we were lucky. In some parts of the Galaxy, few stars would be visible at night, and the absorption of sunlight might have made the formation of life impossible.

Now ponder the effect of the interstellar medium on a spacecraft. Project Daedalus's planners realized that hitting an object only a fraction of a kilogram in size while moving at 12 percent of the speed of light could destroy their vehicle. Even simple protons and electrons encountered at a significant percentage of the speed of light will cause heating and the emission of bremsstrahlung radiation (the radiation that charged particles emit when suddenly slowed or deflected), forcing us to shield sensitive electronics. But dust is more dangerous still, creating temperatures high enough to vaporize some materials. The Project Daedalus final report advocated using a payload shield made of beryllium, a shield 9 millimeters thick with a radius of 32 meters, to protect the vehicle. With a mass of 50 tons, the shield represents only a tenth of the huge Daedalus payload.

The Daedalus final report gives us an idea how such shielding would work. After a journey of fifty years, the stellar encounter would be brief; Daedalus would be within 50 astronomical units of Barnard's Star for just over one hundred hours. But flyby missions like Daedalus have one other significant issue to cope with: Because they do not slow down upon arrival, they encounter the destination solar system, with its far higher density of gas and dust particles, at cruise speed. This means an additional layer of protection for the stellar encounter has to be deployed.

One proposal is to place a shield of dust in the form of a cloud several hundred kilometers ahead of the spacecraft. Any larger objects passing through the cloud would be heated and vaporized before they could damage the payload. To create the cloud, Daedalus designer Alan Bond suggested a small secondary vehicle flying ahead of the main probe whose job is to dispense the cloud material. Alternatively, a particle projector could eject the cloud from the main vehicle, using some form of laser system. Smaller clouds would have to be deployed around the probes Daedalus would release as it entered the Barnard's Star system, although these shields would be allowed to dissipate during the few hours available for planetary observation.

The biggest problem presented by the average, tiny dust grains of interstellar space may well be erosion over the course of a fifty- or one-hundred-year flight. But if interstellar space seems an unlikely place for larger collisions, bear in mind that all it takes is a single pebble-sized object to destroy a spacecraft moving at these speeds. In their book *The Starflight Handbook*, Gregory Matloff and Eugene Mallove have therefore suggested that, in addition to the passive protection provided by a forward shield of the kind used in Project Daedalus, a radar system should actively search for larger particles in the ship's path of flight. A high-power beamed-energy device could then be used to deflect or destroy the object.

As we resolve these issues with precursor missions, we can go about choosing our ultimate targets. In the end, where we send the first interstellar probe will be determined by what it is we want to accomplish. It is all but inconceivable that finding an Earth-like planet circling a nearby star would not move that star to the top of our list for possible missions, meaning that the search for other life in the universe is likely to remain our primary objective. The earliest interstellar missions, probing into the Kuiper Belt and beyond, will teach us enough about interstellar gas and dust both within and outside the heliosphere to make the more ambitious mission possible. These early missions will fill in the gaps, allowing us to build a probe that can survive the journey. We will want to be ready when a green and blue world swims into focus and we learn where we need to go.

CHAPTER 3

THE TROUBLE WITH ROCKETS

George Pal's *Destination Moon* was, in its day, the most realistic science fiction film ever made. Upon its release in 1950, Pal labeled the work "a documentary of the near future," and included within it astronomical sequences by Chesley Bonestell, the artist whose haunting celestial landscapes were the precursors to the images returned by Mariner and Voyager. Science fiction legend Robert Heinlein was a co-author of the script, which was (loosely, to be sure) based on his novel *Rocketship Galileo*. Among the Academy Award-winning special effects is a cartoon sequence featuring Woody Woodpecker.

Pal had folded a complete instructional film into the larger movie, educating his audience about how rockets work, and why they need no air to push against in space. That Pal felt the need to include this basic lesson in physics was a sign of how misunderstood the principles of rocketry remained 24 years after Robert Goddard launched the first liquid-fueled rocket. The *New York Times* had famously attacked Robert Goddard's work on rocketry thus: "Professor Goddard does not know the relation of action to reaction, and for the need to have something better than a vacuum to push on. He seems to lack the knowledge ladled out daily in our schools." That infamous 1920 editorial was not retracted until Apollo 11 lifted off for the Moon.

Pal would go on to make such popular films as *The Time Machine*, *The War of the Worlds*, and *When Worlds Collide*, but interstellar flight never became his subject. It is interesting that there have been no similarly realistic science fiction films about the first journey to another star, no "Destination: Alpha Centauri" or "Destination: Epsilon Eridani." Perhaps that's because the parameters of such a journey are so inconceivably difficult that such a film is beyond the reach of most directors. Not that Hollywood ignores the stars. One of the signal achievements of the original *Star Trek* was that it rarely referred to Earth at all, assuming a space-faring humanity that moved at will among various stellar systems. Better, then, to *assume* when it comes to star travel—it's easier, certainly less tedious for the audience, and after all, we are far from deciding what kind of propulsion system will make star flight possible.

There is no shortage of candidates. In fact, there are so many propulsion systems under such active study that decades of argument and experiment will be needed to sort them out.

We begin with rockets because, until recently, it has been inconceivable to think of space travel without them. Based on twelfth-century Chinese technology, rockets trade off Newton's third law of motion: For every action there is an opposite and equal reaction. The principle is simplicity itself: Stand on a low-friction surface like ice and try to play catch. Every time you throw the ball, you find yourself sliding in the direction opposite the throw. In the same way, push something out a rocket nozzle at high speeds and the rocket moves in the opposite direction. The faster you can push the material, the better the rocket will function. A crucial corollary is that the speed of the material coming out the exhaust depends on its temperature.

Liquid fuel propellants were a true breakthrough compared to those ancient Chinese devices, powered by no more than gunpowder. It would not be until 1926, the day Goddard's rocket lifted off from a snowy farmer's field outside Auburn, Massachusetts, that liquid fuels would come into their own. Goddard's rocket rose to an altitude much less than the length of a Saturn V (a mere 41 feet, in fact), but the concept of mixing combustible liquids—gasoline and liquid oxygen in this case—to feed an engine here took physical form and led to today's boosters, from Apollo to Russia's Proton. Goddard drew on work by Hermann Oberth in Germany and Konstantin Tsiolkovsky in Russia, a reminder that rocketry, like any science, is a kind of edifice, an intellectual ziggurat that rises only as a collection of thousands of bricks.

Goddard had his own dreams, climbing into a cherry tree as a boy to imagine a craft that could reach Mars. He could close his eyes and almost see it being launched from the meadow that spread out beneath his feet. But the man would grow up to be a scientist, and it was a far more practical Goddard, subsidized by grants from the likes of the Daniel Guggenheim Foundation, who wrote a document with the practical goal of prying money out of the Smithsonian Institution. Called "A Method of Reaching Extreme Altitudes" when published in 1920, the report outlined the principles of rocketry as they applied to probing the upper atmosphere with weather recording instruments.

Goddard's report discusses the physics of rocket propulsion and explains how he had used the money so far advanced him by the Smithsonian. For the most part, the report is factual and pitched to a utilitarian audience who wants to know where their money is going. But the young inventor saw a clear track for the future of his work. Saving the kicker for the end, he proposed hitting the moon with a rocket and exploding a charge large enough to be seen from Earth. That led to ridicule, even as enthusiastic work advanced on the other side of the Atlantic. The Verein für Raumschiffahrt (Society for Space Travel) fired its own liquid-fueled rocket on a test stand in July of 1930. The work of its inspired amateurs would eventually be seized by the Gestapo and turned into the first ballistic missile.

The familiar sight of a Space Shuttle liftoff and the similar launches of Delta rockets carrying unmanned payloads, or the French Ariane flaming into the night skies over French Guiana, have fixed an association between rocketry and space in the public imagination. But for all their apparent power, the Space Shuttle's chemical engines push out an exhaust at no more than 10,000 miles per hour, a paltry figure when it comes to interstellar distances. And at a temperature of roughly 5,000 degrees Fahrenheit, the Shuttle is burning its fuel about as hot as engineers can permit without destroying the engine. And there is another problem for chemical fuels: The percentage of the craft that is dedicated to fuel becomes huge. Ninety percent of the weight of the Space Shuttle is propellant. Start talking about missions beyond Earth's orbit and that percentage goes even higher.

Let's consider some numbers. In a typical car, the ratio of fuel to total mass is about 1 to 30, a 3,000-pound car carrying 100 pounds of fuel. To reach the Moon, considerably more of the vehicle must be dedicated to propellant. The Apollo missions demanded a mass ratio (the weight of the rocket when fully fueled compared to the same rocket without fuel) of 600 to 1. As we saw in Chapter 1, to send a chemical rocket on a fast mission to Alpha Centauri would require an amount of propellant larger than all the mass in the observable universe! Even a far more efficient ion engine would need more than 500 propellant tanks the size of supertankers to complete an Alpha Centauri flyby within a century, according to NASA physicist Marc Millis.

And the problems are only beginning. Slowing down takes as much propellant as speeding up; we must, therefore, push that much more fuel. If we wanted our spacecraft to stop once it reached Alpha Centauri, those five hundred supertankers would need to be supplemented by another three hundred million supertankers to make the 100-year journey and stop! As you keep adding more propellant to push still more propellant, the ratio of fuel to payload simply goes off the chart, an unfortunate fact that grows out of the equations that relate mass to velocity change. This is why we use multi-stage vehicles to get into space. As each burned-out stage separates and falls away, the mass that needs to be accelerated by the next stage is correspondingly decreased.

While the power of a thundering Space Shuttle is impressive, we should be more concerned with the efficiency of its engines. In rockets, efficiency is measured by "specific impulse," which measures how much thrust is produced per unit of fuel in each second of the rocket's burn. Put another way, specific impulse measures how many seconds one pound of propellant can produce one pound of thrust. As specific impulse (stated in seconds) rises, it takes less fuel to produce a given amount of thrust, and the amount of payload compared to propellant also rises. Now we can measure how efficient our rockets really are. A typical satellite launch involves a rocket with a specific impulse in the range of 300 seconds. The Space Shuttle's engines, burning a mixture of liquid hydrogen and liquid oxygen, reach a specific impulse of 465 seconds, a good deal more efficient, and more or less state of the art for chemical rockets.

But surely we can do better. For missions into deep space, alternative systems that can achieve a higher specific impulse are the Holy Grail. The October 1998 launch of the Deep Space 1 spacecraft is an early step in that direction. For its fuel, Deep Space 1 used an ion propulsion system that released xenon gas into a thrust chamber surrounded by magnets. When electrons bombard the xenon, the gas is ionized and expelled from the back of the chamber by electrically charged grids at 67,000 miles per hour. The exhaust particles are accelerated not by heat, but by electrical energy.

Deep Space 1 would observe an asteroid called Braille and a comet named Borelly during its lifetime, but the real purpose of the mission was to test new technologies. The ion engine produces only one-fiftieth of a pound of thrust at full throttle, but its specific impulse is high. Compared to a chemical engine, the ion engine yields ten times the thrust for the same amount of fuel. For deep space, the weak acceleration is reminiscent of solar sail and lightsail concepts—even low thrust can achieve high speeds as long as you keep pushing.

And it works. By August of 1999, the spacecraft was experiencing so gentle a push that its speed was changing at a mere fifteen miles per hour each day. But the 9-month testing phase of the mission saw a total of 2.5 months of continuous thrusting, a cumulative effect that changed Deep Space 1's speed by over 1,500 miles per hour. During that time, the engine consumed only 11.5 kilograms (25 pounds) of xenon as it moved over 600 million kilometers (373 million miles) from Earth for its asteroid encounter. By the time the spacecraft was shut down in December of 2001, the ion propulsion system had accumulated 677 days of operation, a dramatic proof of concept for an engine that offers a new versatility for interplanetary missions.

So great are the advantages of ion propulsion that the Jet Propulsion Laboratory considered it a finalist for the Thousand Astronomical Unit (TAU) mission. In one concept, TAU would have used a nuclear reactor to create the needed electricity to ionize the xenon. With a ten-year boost phase, the probe would have reached speeds close to 100 kilometers per second.

But NASA continues to work on a host of propulsion technologies to push rocket efficiency further. One, called the Variable Specific Impulse Magnetoplasma Rocket (VASIMR), heats its fuel with radio waves and uses magnetic fields to shape a stream of hot ionized gases, creating exhaust speeds up to 650,000 MPH. Another employs microwaves to heat two hydrogen isotopes—deuterium and tritium—to temperatures of 600 million degrees Kelvin to ignite fusion. The hellish storm of helium nuclei thus generated could be channeled by magnetic fields to produce thrust. Deep Space 1's ion engine is also being tuned up to increase its specific impulse and triple its power. A fourth concept, the Magnetoplasma Dynamic Thruster, is an electromagnetic engine that offers a new order of efficiency and power to accelerate a plasma exhaust.

All these systems are competing approaches to doing advanced electric propulsion, and all demand robust sources of power aboard the spacecraft, says NASA's Les Johnson. Each tries to take advantage of that power to generate magnetic and electric fields strong enough to accelerate ions and thereby produce thrust. Johnson, who is manager of the In-Space Propulsion Technology Project Office at NASA's Marshall Space Flight Center, was the agency's first manager of its Interstellar Propulsion Project, an effort that has since been rolled into the agency's In-Space effort. The In-Space project has gone on to spawn research into the technologies for interstellar precursor missions to the edge of the solar system and beyond. "We're looking at what I call 'primary propulsion,'" Johnson added, "which means something that gives you a big delta v [change of velocity]. A propulsion system doesn't have to be high thrust; in fact, most of our stuff is low thrust. We've learned that for long voyages, efficiency and low thrust are king. So our charter is to make it more affordable, or quicker, or to increase the payload mass fraction. You want to have the flexibility when you're planning these missions to optimize the science and trip times and costs. And right now chemical propulsion is pretty non-optimal."

It was a similar concern to go beyond the limitations of chemical rockets that drove several projects developed by the U.S. Atomic Energy Commission following World War II. The idea was to find ways to exploit the energy locked up inside the Einsteinian equations relating energy to mass; nuclear fuels, after all, contain vast amounts of energy per unit mass compared to the best chemical fuel, as much as a million times more. The projects involved building nuclear engines, and one—Nuclear Engine for Rocket Vehicle Application, or NERVA—came close to producing a flight prototype before the program was cancelled in 1972. The science fiction concept of a "nuclear drive" was rarely explained in the pages of the magazines that published stories about spaceships using nuclear power, but the reality of nuclear engines would probably have struck many of their readers as pedestrian. The nuclear reactor on board would essentially be used to supply heat, a process called "nuclear-thermal." A propellant like hydrogen would be pumped through the reactor and then channeled through an exhaust nozzle as it expanded, thus generating thrust.

NERVA actually began with an earlier program at Los Alamos called Rover, created in 1956. Work on the combined program from 1956 until 1971, twenty-three tests in all, produced specific impulses of 850 seconds, almost twice that of the Space Shuttle's main engines, and exhaust temperatures of 5,500 degrees Fahrenheit. NERVA used a solid graphite core, but the Rover program also investigated using a gaseous nuclear fuel, which would allow operating temperatures in the tens of thousands of degrees, and specific impulses as high as 5,000 seconds. If successfully developed, a gas-core nuclear rocket would allow a Mars mission to be flown with a three-month transit time to Mars and a four-month return voyage.

Nuclear-electric methods are also an option, using a nuclear reactor as an electrical power source that, like an ion drive, accelerates the exhaust propellant to a high velocity. Heat problems are serious in such designs, because a working reactor can only be allowed to reach temperatures that the spacecraft's structure can tolerate. That limited early reactors to maximum temperatures in the range of 4,500 degrees Fahrenheit, and the engines were massive.

Nuclear rocket propulsion re-emerged as a serious research topic in the 1980's as part of the Strategic Defense Initiative. One project, called Timberwind, produced no full-power engine, although various components were tested. Much of the work performed during the decade was declassified in 1991, and the technology evaluated for possible use in interplanetary missions. Timberwind focused on a so-called particle-bed reactor (PBR), with hydrogen injected into a reactor core of spherical fuel particles. A key advantage of the technology was that the nuclear thermal engine could be made smaller and lighter than designs like NERVA. Instead of tons, PBR systems weighed hundreds of pounds. One design that followed the PBR program—called MITEE, for Miniature Reactor Engine—is a nuclear engine with a mass of only 140 kilograms.

MITEE's designer, Plus Ultra Technologies of Stony Brook, New York, envisions using MITEE for missions outside the solar system, generating velocities of up to 40 AU a year by harnessing a gravitational assist by the Sun through an extremely close pass. Gravity assists around the planets have already been demonstrated by spacecraft like the Voyagers and Galileo. In each case, the spacecraft is flung as if from a slingshot into a new and much faster trajectory. The Kuiper Belt would be in reach for such a vessel, and perhaps even the gravitational lensing point some 550 AU out, where the Sun's gravity bends and amplifies the light of distant objects, making this a prime target for future astronomy.

One sign that work on nuclear propulsion is still active is Project Prometheus, a nuclear power scheme being developed by NASA and the Department of Energy. As with other nuclear propulsion concepts, weight is a key factor—nuclear reactors are not light, and every ounce that can be saved on a space mission increases the available payload. The Prometheus idea is to create compact reactors to fly on a new generation of robotic space probes, including the Jupiter Icy Moons Orbiter (JIMO), which would depart no earlier than 2012 and rendezvous with Europa, Ganymede, and Callisto. JIMO has reached the stage of study contracts being awarded to Boeing, Lockheed Martin, and Northrop Grumman Space Technologies. Developing radiation shielding for JIMO's science instruments is one of many issues now under discussion, as is radiating waste reactor heat into space.

JIMO should be able to provide more power than any previous spacecraft has enjoyed. Current probes, like the Galileo spacecraft that orbited Jupiter between 1995 and 2003, use radio-thermal isotope generators to produce electricity aboard the vehicle, drawing their power from plutonium. Prometheus is an attempt to significantly improve the power supply aboard such vehicles while creating a generation of nuclear electric thrusters—nuclear reactors coupled to ion drive engines—that can propel the probe on the seven-year journey to Jupiter. No probe has ever had the internal power resources Prometheus offers, with huge benefits for instrumentation. "For thirty years we've been working with a few light bulbs," says Colleen Hartman, who directs NASA's efforts at solar system exploration. "All of a sudden, this is stadium lighting, kilowatts instead of watts."

Yet all these designs lag well behind the dreams of interstellar theorists as they were shaped in scientific debates and barroom bull sessions in the 1960s and 1970s. The nuclear constraint is one reason: Policy makers frown on nuclear technologies flying on spacecraft because of the risk of radioactive debris, and every launch failure or Shuttle disaster only drives home the apprehension. Project Prometheus was met with protesters outside the Albuquerque venue of the Space Technology & Applications International Forum when it was announced in February of 2003 at a symposium on space-based nuclear power. Anti-nuclear groups also opposed Cassini, a probe to Saturn and its moon Titan, because the spacecraft relied on plutonium as the power source for its instruments. There are no signs that anti-nuclear sentiment will ease any time soon, despite our continuing reliance on fossil fuels whose polluting effects are arguably worse than any nuclear spills. And the nuclear dilemma has already played a role in squashing what had appeared to be the boldest and most innovative of all propulsion designs of its day, a scheme to drive a spacecraft by a series of nuclear explosions.

That project was called Orion. Conventional nuclear drives worked by heating a propellant, expanding it, and channeling its exhaust stream for thrust. Project Orion was developed along different lines. The idea was to explode a nuclear bomb outside the rocket, but close enough to a flat pusher plate so as to drive it forward by the transfer of momentum. The explosions would vaporize an inert propellant between the pusher plate and the bomb. The plate itself would be attached to the spacecraft by what surely would be ranked as the world's most efficient shock absorber.

Orion grew out of work performed by Stanislaw Ulam and Cornelius Everett at the Los Alamos Scientific Laboratory and continued through experiments at the Eniwetok nuclear test facility in the Marshall Islands of the Pacific. It proved to be a remarkable fact that ordinary metals could withstand the high surface temperatures of such explosions. By 1958, when the General Atomic division of the General Dynamics Corporation began active work on what was now being called nuclear pulse propulsion, the concept seemed practical and, despite the inevitable bureaucratic wrangling caused by multiple overseeing by the Advanced Research Project Agency, the Air Force, and NASA, the Orion team was able to produce what seems to have been an efficient design using a dual shock-absorber system. A miniature Orion model using an aluminum pusher plate and chemical explosives was successfully flown in late 1959, rising 100 meters and demonstrating that a vehicle powered by pulse propulsion could be stable in flight.

Contrast Orion to a conventional rocket. The Space Shuttle's propellant, carried on board, is ignited and driven through a nozzle. Orion used plutonium for fuel, but the propellant was actually whatever material would be vaporized between the exploding bombs and the pusher plate. Anything that would vaporize would do; the Orion team originally chose plastic, later deciding to combine propellant and bomb in a single unit. Freeman Dyson talked about re-fueling Orion at Saturn's moon Enceladus, which was believed to be rich in ice and hydrocarbons. Indeed, an Orion expedition could pick up propellant almost anywhere it journeyed, from ice on frozen Europa to lunar regolith from Mare Crisium.

Because a chemical rocket ignites fuel inside the vehicle, temperature becomes a key limit on how powerful the engine can be. Orion reaches far higher temperatures, but they are quickly dissipated. The plasma from the explosion and the vaporization of the propellant would hit the pusher plate a hundred times faster than the speed of a conventional rocket's exhaust. That produces temperatures up to 120,000 degrees Fahrenheit, but because the plasma is only in contact with the plate for the tiniest fraction of a second, the heat is survivable. The blast pushes the ship with a theoretical specific impulse ranging from ten thousand to one million seconds, a value that dwarfs anything available from competing designs.

Wedding high specific impulse with high thrust was what made Orion unique. Deep Space 1's ion drive offers high specific impulse, but its thrust is low. A chemical rocket offers high thrust but low specific impulse. Orion could deliver both, with the dazzling benefit that the bigger you made the spacecraft, the more impressive its efficiency. Indeed, one of the many beauties of the Orion design was the size of the resulting vehicle. In the era of cramped Mercury and Gemini capsules, Orion promised enormous vessels carrying a crew of one hundred or more, launched directly from Earth and venturing not only to the Moon but to Mars and beyond. A later design scaled the vehicle back to allow Saturn V rockets to boost the disassembled spacecraft into orbit as an upper stage, with final construction in orbit.

Even this restricted design offered a startling mission scenario: an eight-man Mars mission with one hundred tons of supplies making the journey to the Red Planet in 125 days. Intriguingly, weapons designer Theodore Taylor, who became Orion's guiding force in the era of its development, modeled the project's management after Germany's Society for Space Travel, among whose members was Wernher Von Braun, who became an enthusiastic supporter of Orion. The idea: hold down the managerial overhead and minimize bureaucracy, an ironic goal given Orion's troubles in trying to navigate through the shoals of three federal agencies. The signing of the nuclear test ban treaty in 1963 was but one of a series of blows to the project, which eventually lost its funding after expending $11 million over the seven-year course of its life. Even today, much of this work remains classified, and various parts of the Orion technology went on to play a role in the development of the Strategic Defense Initiative in the 1980s.

Theoretical physicist Freeman Dyson, a major player in the project, went on to push the design to its limits, pondering using fusion explosions to create even more powerful vehicles. In one Dyson version, the ship would use a copper pusher plate twenty kilometers in diameter. Fifteen billion kilograms of deuterium (an isotope of hydrogen with a nucleus of one proton and one neutron, double the mass of the ordinary hydrogen nucleus) would go into the production of some thirty million nuclear bombs, each of which would explode 120 kilometers behind the vehicle at intervals of 1,000 seconds. With a total acceleration time of five hundred years—and a comparable time for deceleration—this mammoth super-Orion would carry a colony of 20,000 Earth people to Alpha Centauri. Flight time: 1,800 years, making it a true multi-generation ship, where the distant descendants of the initial crew arrive at the target to make a new start for humanity. A second Dyson design was leaner in concept, using 300,000 bombs to reach a final velocity of 10,000 kilometers per second, with arrival at Alpha Centauri in 130 years.

Although Dyson gave interstellar concepts a serious look, his interest was largely theoretical; he never saw Orion as practical for anything more than interplanetary travel. But the Orion concept remains fertile, and as upgraded by Dyson with fusion bombs, it is a propulsion system that could be built without violating any known laws of physics. It is also the only viable concept for launching large payloads on interstellar missions without inventing entirely new technologies. Even today, a descendant of the nuclear pulse propulsion system of Orion, now labeled External Pulsed Plasma Propulsion (EPPP), is under consideration for missions to the outer planets, as well as for defense against Earth-crossing comets or asteroids.

Science fiction makes sporadic use of nuclear pulse propulsion—perhaps this is a case where the reality seems more implausible than the fiction that could be spun out of it. Who, after all, would imagine a crew surviving a voyage sitting behind a pusher plate driven by the most violent release of energy scientists know how to create? One notable exception is *Footfall*, a novel by Larry Niven and Jerry Pournelle, in which humanity uses an Orion-like spacecraft as part of its defense against alien invaders. Both Poul Anderson and Stephen Baxter have worked entertainingly with the idea. And Vernor Vinge offered a truly futuristic take on Orion in his *Marooned in Realtime*, which envisions a spacecraft encased in a spherical time-stasis field, after which it is safe to explode a series of atomic devices next to it for propulsion. An Orion-style craft even shows up in Hollywood's *Deep Impact* (1998), in which astronauts make a cometary rendezvous to save Earth from catastrophe. But the film's special effects did not include a realistic Orion, depicting a vehicle without noticeable pulsed propulsion. Theodore Taylor once described how an actual Orion vehicle would look to Earthbound observers as it made its way to space, a far cry from *Deep Impact*: "Most of what you'd see would be images, not of the explosions themselves, but of their effects on the upper atmosphere—flashes of multicolored light." It might be possible to see "the expanding explosion of each pulse, something like a bright cone. It might even be blinding."

Such demanding special effects may have been what dissuaded Stanley Kubrick from using a pulsed-nuclear drive for his Jupiter-bound spacecraft in *2001: A Space Odyssey*. An early version of the screenplay, which did use the Orion concept, paints the scene as the spaceship *Discovery* leaves the vicinity of Earth to begin its mission to Jupiter: "*Discovery* 1,000,000 miles from earth. See Earth and Moon small. We see a blinding flash every 5 seconds from its nuclear pulse propulsion. It strikes against the ship's thick ablative tail plate."

Long after the demise of Orion, engineer Robert Zubrin, perhaps best known for his tireless advocacy on behalf of the Mars Society, created an offshoot of the Orion idea called the Nuclear Salt Water Rocket. Considering its uses in a hypothetical mission to Saturn's moon Titan, Zubrin called for a fuel of 20 percent enriched uranium deployed in the form of a soluble salt that would be dissolved in ordinary water. The concentration of fissionable material is such that the fuel must be stored in a neutron-absorbing fuel tank that holds the fuel stable while it is waiting for use. The actual fission reaction begins inside a reaction chamber, producing a continuous nuclear explosion, but a controlled one whose energy peaks just outside and behind the spacecraft. A key problem remains the highly radioactive exhaust, and the difficulty in designing an exhaust nozzle that could handle the extreme heat.

Zubrin's second mission concept takes us interstellar. For a 120-year mission to Alpha Centauri, his spacecraft would use a more uranium-rich salt water mixture, 2,700 tons of this fuel, to achieve a speed of close to 4 percent of the speed of light. Intriguingly, the fuel for such a mission is plentiful. It is available from decommissioned nuclear weapons left over at the end of the Cold War. Moreover, it does so using existing technologies. As physicist and science fiction novelist John Cramer puts it, "An unmanned Alpha Centauri probe of the type that Zubrin suggests could be built starting today and at a cost that I would guess would be much smaller than the growing price tag of NASA's troubled Space Station Freedom project. The anticipated mission time of 120 years is a long time. But the sooner we start, the sooner we (or our descendants) will get a close-up look at our neighboring star systems."

Both the nuclear salt water rocket and Project Orion's cascade of exploding bombs take advantage of fission—dividing a heavy atomic nucleus like uranium or plutonium into fragments while releasing huge amounts of energy—instead of fusion, which causes lighter elements like hydrogen to fuse into heavier ones like helium, with more powerful energy output. Not long after the demise of Orion, fusion was to have its day, spurred by growing interest in the idea of controlling fusion reactions to produce power. It was in 1966 that the Jet Propulsion Laboratory's Dwain Spencer outlined the principles of a fusion engine designed to burn deuterium and helium-3 (an isotope of helium with a nucleus of two protons and one neutron). Earth-bound testing in today's tokamak nuclear reactors has worked with deuterium in combination with tritium (a hydrogen isotope with two neutrons and a single proton), using powerful magnetic fields to contain the plasma, but the deuterium/tritium combination releases the bulk of its energy in the form of radioactive neutrons, creating a safety issue for any manned spacecraft carrying a nuclear engine. The deuterium and helium-3 combination, on the other hand, produces one-hundredth the amount of neutrons of the deuterium/tritium reaction, so a fusion-based space drive would need less radioactive shielding. Moreover, the output of the deuterium/helium-3 reaction—protons and alpha particles—can be manipulated by a magnetic nozzle, whereas the neutrons produced by the deuterium/tritium reaction cannot. Finally, using deuterium/helium-3 produces a reaction that is uniquely productive in terms of the amount of energy released from a given amount of mass. Only antimatter produces more of a wallop.

The concept of plasma is critical to these fusion concepts. Plasma is often called the fourth state of matter, the others being solid, liquid, and gas. Heat a gas sufficiently and its electrons will break away from the nuclei of their atoms, causing the electrons and nuclei to move independently. Because they are electrically charged, plasma particles can be manipulated by magnetic traps to contain them without coming into contact with the walls of a combustion chamber, and they can be directed out the back of the spacecraft. While atomic nuclei in the other states of matter are positively charged (and therefore repel each other), heating a gas into a plasma puts nuclei into violent motion in a confined space, where collisions between them are likely and fusion possible.

Spencer's 1966 paper "Fusion Propulsion for Interstellar Missions" described an engine that burned deuterium and helium-3 gases in a combustion chamber that was ringed with superconducting magnetic coils to keep the hot plasma from contacting the chamber's walls. The exhaust would be directed out the back of the spacecraft, pushing it to three-fifths of the speed of light at maximum velocity. Given the acceleration and deceleration times needed for the mission, the five-stage spacecraft would take fifty years to make the crossing to the nearest star. Several years later, Rod Hyde and colleagues at Lawrence Livermore Laboratory proposed using deuterium and helium-3 in the form of frozen pellets, exploding hundreds of them every second in an inertial confinement fusion (ICF) reactor whose exhaust would be shaped, as in Spencer's design, by magnetic coils. Drawing power from the current induced in a pickup coil by the fusion reaction, the engine would use lasers to ignite fusion in the pellets.

An interstellar probe built with such an engine would attain one-tenth of light speed for a flyby mission to a nearby star. Soon after Spencer's paper appeared, the possibilities of helium-3 as a fusion fuel were enhanced when Apollo astronauts found the isotope on the Moon in 1969. While only hundreds of pounds of helium-3 can be found on Earth, some scientists believe that a million tons of the stuff are likely to be located on the lunar surface, created by the charged particles of the solar wind as it strikes the surface. "There's enough in the Mare Tranquillitatis alone to last for several hundred years," says Apollo 17 astronaut Harrison Schmitt, who foresees the use of helium-3 as a fuel source for the twenty-first century.

There are vaster storehouses of helium-3 in the solar system, including the atmospheres of gas giant planets like Jupiter and Saturn. When Alan Bond proposed that the British Interplanetary Society begin designing the Project Daedalus starship in 1972, the six committees given the task ruled out all propulsion methods but pulsed fusion, and settled upon fuel in the form of frozen pellets of deuterium and helium-3. The pellets were to be bombarded, one pellet at a time, in the combustion chamber by electron beams, creating a fusion "micro-explosion." Using deuterium and tritium would have allowed for a fusion reaction that is easier to initiate, but the Project Daedalus team, like Spencer, saw that deuterium and helium-3 provided a more manageable exhaust at greater power.

Drawing on a key paper by Friedwardt Winterberg and on newly declassified work on initiating fusion in small amounts of high density materials with lasers, the team concluded that their design was a reasonable extrapolation of existing technology, if an enormous one—Project Daedalus' early designs called for a reaction chamber 330 feet in diameter. Later designs reworked Daedalus into a two-stage mission, burning the first stage for two years and the second for almost as long, dropping fuel tanks as they were exhausted. Having used up the last of its fusion pellets, the probe would coast at 12 percent of the speed of light to its rendezvous with Barnard's Star.

The Daedalus mission would require fifty billion fuel pellets, the force of whose explosions was to be channeled in the form of plasma out the rear of the spacecraft. As for helium-3, the project estimated a need for 30,000 tons of the stuff, along with 20,000 tons of deuterium. The Daedalus team studied several alternatives for propellant acquisition, including collecting helium-3 ions from the solar wind, but settled on mining it from the helium-rich atmosphere of Jupiter. As outlined in the project's final report, floating factories would be placed in the Jovian atmosphere, serviced by orbital transports and using waste heat from their power reactors to generate lift. A large number of these "aerostat" factories would be built, separating and liquefying the isotopes needed.

The huge infrastructure demands of interstellar flight seem to await us no matter which propulsion system we use. What Daedalus demanded was nothing less than the industrialization of the solar system, building the base from which the interstellar probe could be constructed. As the final report put it, "... an undertaking on the scale of Daedalus fits naturally into the context of a solar system-wide society making intelligent use of its resources, rather than a heroic effort on the part of a planet-bound civilization." And it presupposes major advances in the technology of inertial confinement fusion. In this system, a spherical fuel pellet is compressed by hitting it from all sides by powerful laser beams, compressing the sphere to make it dense enough—and hot enough—to light the fusion reaction. Igniting a sustainable reaction remains the most significant problem for all fusion propulsion concepts. And it is a particularly complex problem for inertial confinement systems because ICF attempts to compress fusion fuel to high densities while simultaneously delivering fusion energy to the engine. That calls for a tight balancing act between compression and ignition that has bedeviled attempts to perfect ICF systems. "On paper, ICF is reasonable," says Terry Kammash. "But in practice, it has proven a tough challenge. Very tough."

Kammash, a professor at the University of Michigan's Department of Nuclear Engineering and Radiological Sciences, has proposed a variant approach to igniting fusion called Magnetically Insulated Inertial Confinement Fusion (MICF). Significantly, MICF does not depend upon compression as part of the ignition process. Instead, it uses lasers or particle beams to heat centimeter-size fuel pellets made of tungsten or gold, their inner walls coated with fusion fuel. The lasers fire into a small hole in each pellet to strike its inner wall. A plasma is thus created which becomes hot enough to ignite the fusion reaction. A beneficial side effect is that powerful magnetic fields are induced in the plasma as fusion is ignited, slowing down the flow of heat from the hot plasma to the wall. Acting as heat insulators, these magnetic fields allow the reaction to burn long enough to produce more energy than it absorbs. An MICF system developed as a spacecraft engine would require a mammoth laser, but Kammash went on to suggest replacing the laser with the most exotic material known on Earth: antimatter.

Could antimatter be the key to sustainable fusion, and if so, how would we use it without it meeting normal matter and thereby annihilating the spacecraft? The questions are countless, but remarkably, this staple of science fiction has now entered the realm of serious propulsion study, though in a form much different from the Galaxy-spanning engines familiar to Star Trek fans. The Enterprise may have used antimatter in its warp drives, but as we'll see in the next chapter, huge issues in producing antimatter and channeling its almost inconceivable energies remain to be solved. Even so, we know how to make antimatter and we make it today, if only in the tiniest of quantities, and we're beginning to learn how to extract enough of its energies to make some forms of fusion engines possible.

CHAPTER 4

THE ANTIMATTER ALTERNATIVE

Science fiction's treatments of antimatter for propulsion have sometimes had the feel of a conjuring trick, where the illusionist distracts the eye while doing some completely obvious thing to make the trick work (don't blink or the Statue of Liberty will disappear). Bypassing vast technical problems to create an easy route to the stars, such stories use antimatter the way some writers used hyperspace—as a plot device without serious foundation in science. But the best science fiction demands more of itself, in keeping with the field's premise that good stories come only from science that is both believable and consistent. Such work examines a theory from every direction, probing not only for drama but for what Richard Feynman wonderfully called "the kick in the discovery." Both laboratory researchers and fictional spacemen can relate to that, but the discovery had better be, if not real, at least plausible.

And that has put the creators of fictional antimatter starships in a quandary. Because the plain fact is that, other than presuming vast amounts of matter and antimatter somehow annihilating themselves in a combustion chamber the likes of which is currently unimaginable, we have no idea how to build a true antimatter engine. Writers will have their hands full if they want to sketch out a scientifically accurate future, for nobody is sure (1) how we'll begin to produce enough antimatter to do such a thing; (2) how we'll transport and contain that antimatter fuel so it doesn't annihilate itself at the wrong moment; (3) how we'll channel the antimatter through a propulsion system to get the maximum thrust out the back of the engine.

For now, those writing about antimatter starships can do little more than suggest engines with plenty of punch and refer to the antimatter that fuels them. But the good news is that practical propulsion concepts are being devised, storage and transportation issues are being resolved, and there is hope of increasing our currently minuscule antimatter production rate within the not-so-distant future.

It is not too much to say that antimatter research totters on the edge of breakthrough. More than plausible, antimatter is real, is being produced, and is under active investigation. In fact, Earth-bound applications of antimatter now range from the treatment of cancer through antiproton irradiation to positron emission tomography (PET), an imaging system that has helped doctors diagnose disease by examining body tissues. Antimatter's potential as a rocket fuel is immense because it can, theoretically, convert one hundred percent of its mass into energy. (By contrast, a fission reaction like an atomic blast uses about 1 percent of the energy locked up inside matter.) Of course, channeling the energy of antimatter into a thruster would be extremely difficult. The most realistic scenarios for using antimatter in rockets involve tapping small amounts to power fusion reactions, at least until we know how to make it in quantity. As we'll see, the possibilities are breathtaking.

Antimatter has a long and colorful history both in science and science fiction. It was in 1928 that British physicist P. A. M. Dirac brought Einstein's special theory of relativity to bear on the behavior of electrons in magnetic and electric fields. Dirac worked just twenty-three years after the publication of that theory, which explained the relationship between mass and energy and explored the dynamic connections between space and time. And he also worked in the shadow of Max Planck, who had shown that light must come in packets rather than waves or particles. That work would lead directly to the creation of quantum theory, which could describe slow-moving particles but broke down when predicting their behavior when they were moving close to the speed of light.

When Dirac combined quantum theory with special relativity, he created equations that could explain the behavior of such particles, but which also produced an unexpected result: There should exist an antiparticle to the electron—that is, a particle of the same mass as the electron but housing a positive electric charge as opposed to the normal negative electric charge. Carl Anderson discovered the antiparticle in his laboratory at the California Institute of Technology in 1932, and the editor of the journal that published his work gave the particle the name "positron." It followed that if electrons had antiparticles, so did protons, and in 1955 the first observed antiproton was created at the Berkeley Bevatron, a machine that drove protons into collisions at high speeds. The antineutron was discovered the following year.

What is odd is that antimatter shows up so rarely in nature, appearing only under extraordinary circumstances. Recent NASA research, for example, indicates that a solar flare in July of 2002 produced about a pound of antimatter. Cosmologists believe that a slight imbalance in the early universe between protons and antiprotons led to the annihilation of all but the surplus protons. In one of his "Alternate View" columns in *Analog Science Fiction and Fact,* physicist John Cramer pegs the early universe as having roughly 100,000,001 protons for every 100,000,000 antiprotons, seeing the electrons and protons that comprise the visible universe as "... the few ragged survivors of the 'antimatter wars' of 16 billion years ago."

In the absence of antimatter around us, however, we have learned to create it by smashing protons moving at extreme speeds into targets inside particle accelerators. Lined with superconducting magnets, accelerators like those at Fermilab (the Fermi National Accelerator Laboratory in Batavia, Illinois) and CERN (the European Organization for Nuclear Research) give the protons enough kinetic energy to turn them into a shower of particles and antiparticles, along with x rays and gamma rays, at the blinding moment of collision.

What makes antimatter attractive as a propulsion candidate is that antimatter particles—antiprotons and antielectrons (positrons) —react violently when they meet normal matter. Physicists call the process "annihilation." We know today that a single gram of matter meeting a gram of antimatter would release the energy of a 20-kiloton bomb, the size of the Hiroshima weapon. No wonder antimatter so quickly caught the eye of rocket scientists. The earliest studies of antimatter rocketry were conducted in the 1950s by a German researcher named Eugen Sänger, whose "photon rocket" design, presented in a lecture at the Fourth International Astronautical Congress and subsequently published, would use the gamma rays produced by the annihilation of electrons and positrons for thrust.

The problem was that controlling the exhaust seemed impossible. Gamma rays carry so much energy that they can penetrate any known material, and all Sänger's attempts to find a way to direct the exhaust stream failed. It wasn't the first failure for Sänger, whose 1935 design for an "antipodal bomber" visualized a world-circling craft that could bomb New York, flying at Mach 10 and reaching an altitude of 175 miles. Although static engine tests were conducted, the antipodal bomber was put aside so that the German rocketry community could concentrate on the V-2.

But if a photon rocket wouldn't fly, the discovery of antiprotons not long after Sänger's lecture would make a different form of antimatter rocket more feasible. When antiprotons and protons meet, they release vast amounts of energy given off not only as gamma rays but also as pi-mesons, short-lived particles also known as pions that can be neutral or positively or negatively charged. Because many of the pions are charged when they emerge from the proton/antiproton annihilation, their paths can be curved by sending them through a strong magnetic field. Various methods of channeling charged pions into an exhaust stream have been suggested, including a design by Robert Forward that would use superconducting coils to shape the flow of pions or use the flow to heat a propellant such as liquid hydrogen. In one design by David Morgan of the Lawrence Livermore National Laboratory in California, a 3-meter nozzle uses magnetic coils to channel pions created by the annihilation of antiprotons and neutral hydrogen atoms into an exhaust. The exhaust velocity on the Morgan engine reaches an astonishing 94 percent of the speed of light.

When antimatter meets normal matter, the total energy of their masses is released in the resulting annihilation, the amount being given by Einstein's equation $E = mc^2$. This makes antimatter the most concentrated form of energy we know, allowing theoretical mass ratios—the weight of the fully-fueled spacecraft versus the empty spacecraft— of 5 to 1. Contrast this with the 100 to 1 mass ratio of the Daedalus probe! Robert Forward calculated that for a 1-ton robotic probe moving at one-tenth the speed of light on a scientific mission to Alpha Centauri, an antimatter rocket would require no more than four tons of liquid hydrogen or other propellant, along with some forty pounds of antimatter. If a faster mission were required, doubling the speed would call for the same four tons of liquid hydrogen, but four times the amount of antimatter. A two-stage mission that would decelerate at Alpha Centauri for extended exploration would require 770 pounds of antimatter and 24 tons of liquid hydrogen.

Clearly, a small amount of antimatter would offer the propulsive power of tons of chemical propellant. NASA estimates that a gram of antimatter carries the potential energy of a thousand of the huge external fuel tanks carried by the Space Shuttle. And the matter/antimatter annihilation is one hundred times more efficient than a fusion reaction. The allure of antimatter is offset by the practical difficulties of making enough of it to make such missions possible, finding ways of storing large amounts for missions that might last decades, and channeling that energy into a practical rocket engine. What theory proposes, engineering cannot always deliver, but we'll surely keep trying.

Antimatter's properties were examined from all sides by science fiction writers before they emerged in the more respectable venues of astronautical journals. The potential of the stuff quickly caught the eye of John Campbell, editor of *Astounding Science Fiction* and the man who, a decade later, would publish the groundbreaking article on solar sails, anticipating a propulsion system now under intense scrutiny for interstellar precursor missions. Author Jack Williamson had visited Campbell in the early 1940s and learned from him about Carl Anderson's discovery of the positron. Campbell suggested that Williamson explore the concept of entire worlds based on antimatter, an idea Williamson referred to as "contra-terrene." Williamson's stories on the subject of "CT" appeared in the magazine under the pen name Will Stewart, and were later collected in two volumes, *Seetee Shock* and *Seetee Ship*. Later, the swashbuckling tales would make an appearance before an even larger audience as *Beyond Mars*, a comic strip that Williamson scripted for the New York Sunday News, although the antimatter theme was by then but a minor one.

Like a glittering thread, antimatter runs through the tapestry of science fiction from this era and later, though its implications are only vaguely glimpsed. When Isaac Asimov needed to create the science behind the brains of his robots in his famous series of stories, he chose information pathways that were created by positrons. Gene Roddenberry asked Asimov's permission to make Data a positronic robot in *Star Trek: The Next Generation*, and the *Enterprise's* engines were themselves powered by antimatter. Exactly how these drives worked had to be left to speculation, for we are only now beginning to understand antimatter's properties, not to mention the best ways to control it.

Nonetheless, antimatter is routinely made and manipulated today at places like CERN and Fermilab. Here, a vacuum chamber lined with magnets, vacuum pumps, and high-voltage instruments is the scene. Electric and magnetic fields (with the help of radio waves) push protons to speeds approaching that of light. When the high-energy protons smash into a metal sheet, a menagerie of particles and gamma rays is created, including tiny amounts of antimatter. The antimatter particles—about one antiproton for every million protons that collide with the target—can be separated from conventional matter by magnets because a particle and its antiparticle have opposite charges and curve in different directions in a magnetic field.

What would it take to make antimatter a serious candidate for propulsion? Former Penn State University physics professor Gerald Smith has worked on the confinement of antimatter at CERN and Fermilab, and is now a managing member of a Santa Fe, New Mexico-based antimatter company called Positronics Research LLC, which works with national research centers to improve our production of antimatter and to create practical uses for it. Smith collaborated with physicists Steven Howe, Raymond Lewis, and Kirby Meyer to design AIMStar, Antimatter

Initiated Microfusion Starship, which would fly an interstellar precursor mission.

The goal of the AIMStar mission is to deliver a 220-pound probe a distance of 10,000 AU, over 930 billion miles from the Earth and ten thousand times its distance from the Sun. That puts the spacecraft in the domain of the Oort Cloud, where trillions of comets are thought to congregate in a vast spherical cloud. The flight plan calls for continuous acceleration for four to five years, after which the spacecraft would coast forty-five years before reaching its goal, making a flyby of Pluto in the first part of its journey. The big drawback: producing enough antimatter to make the journey a possibility, not to mention the related problem of building a workable, long-life nuclear-fusion engine. For AIMStar is really a fusion device that uses antimatter to light the reaction.

The AIMStar engine relies on a Penn State discovery first applied to fission: Antiprotons can induce a highly effective form of nuclear fission that allows almost all the energy from the fission reaction to be transferred to the spacecraft's engine, as opposed to simply transferring the heat energy from the uranium core to surrounding chemical propellants. Such an antiproton-catalyzed micro-fission (ACMF) engine could serve for interplanetary missions, reaching Mars in a travel time of about a month.

But faster transit times are needed to reach the Oort Cloud, and AIMStar is designed as a logical extension of the ACMF technologies. For AIMStar, a small amount of fissionable material is used in combination with between 30 and 130 milligrams of antimatter to create a fission reaction whose energy is used to heat a fuel to the point of fusion. The resulting engine is almost five times as efficient as an ACMF device. The AIMStar engine is a practical recognition of the problems physicists have had in trying to create fusion using laser or particle beams, or in containing fusion reactions with intense magnetic fields. By injecting fusion fuel droplets of deuterium and helium-3 into a cloud of antiprotons, AIMStar would light the fire, creating an exhaust stream whose speed might reach 200 million miles per hour.

Is AIMStar aiming too high? The system sounds like a fast track to the outer solar system, but its design is not without significant drawbacks. A major problem is mass: These engines are simply too heavy to fly without sacrificing payload, a fact that has inspired lighter designs we'll examine in a moment. Even so, antimatter-initiated micro-fusion is one step along a roadmap of the sort NASA likes to talk about as leading to its interstellar goals. The AIMStar team thought an interplanetary mission using the somewhat less futuristic ACMF engine would be a logical first step toward proving antimatter propulsion viable. Their Ion Compressed Antimatter Nuclear rocket, called ICAN-II, is the outcome. In one configuration, the design would echo Project Orion in pushing a spacecraft by nuclear explosions, but in this case the fuel would not be the Earth's stockpile of nuclear weapons but some 200,000 six-ounce pellets made of uranium and liquid hydrogen. Using antiprotons to trigger the nuclear reaction, the pellets could be exploded at the rate of one every second to obtain the needed thrust.

The beauty of such a concept is that it makes interplanetary destinations, already reachable through conventional liquid-fuel technologies, available for massively larger payloads. ICAN-II is envisioned as an 800-ton spacecraft, enough to carry substantial scientific instrumentation and a large crew. The imagery of Stanley Kubrick's *2001: A Space Odyssey* comes to mind, the vast spacecraft sliding into orbit around Jupiter as its lone surviving astronaut grimly struggles to resolve the problem of the demented computer. We can hope that ICAN-II will one day operate with a more reliable computer system than HAL, for its vision of huge manned vehicles contrasts wonderfully with today's small, robotic probes.

A key to antimatter propulsion is the development of better ways to contain it. Antimatter is normally stored in so-called Penning traps, which use electrical and magnetic fields to hold the charged particles in suspension from conventional matter. But such traps are bulky and cumbersome, and contain a relatively small number of antiprotons. Penn State researchers have worked to extend the capacity of such traps to hold more and more antiprotons. The Penn State Mark I Penning trap weighs in at just over two hundred pounds and can store 10 billion antiprotons for up to a week.

This work is feeding into NASA studies that would increase these capabilities by 100 times. Marshall Space Flight Center's high-capacity trap called HiPAT (High Performance Antiproton Trap), shown to me that day at Marshall by Ray Lewis, could be wedded with an antimatter plasma gun design that was studied by Penn State for possible use on AIMStar. And there are other ways around the containment problem, such as changing how the spacecraft generates its thrust and reducing the amount of antimatter needed. The idea in all such theoretical work is to tweak one good idea with another until a design emerges that is capable of being tested experimentally.

Which is exactly what is happening in Huntsville, Alabama, where current fusion concepts remain focused on interplanetary possibilities. A team at Marshall Space Flight Center led by William Emrich has been developing a fusion engine for solar system missions. Emrich's plan is to heat deuterium and tritium to some 600 million degrees Kelvin using microwaves. By creating a cylindrical magnetic field, the charged helium nuclei resulting from the fusion can be channeled through a nozzle to produce the needed thrust, up to 300 times the thrust of a conventional chemical rocket engine. The Marshall work contains the plasma by using a "gas dynamic mirror" (GDM), a cylinder wrapped with current-carrying conductors that avoids the instabilities of older containment systems.

A GDM-based engine would allow for a specific impulse in the range of 100,000 seconds, compared to nuclear-thermal rocket designs that weigh in at under 1,000 seconds. But Marshall is also investigating an engine that could theoretically double the GDM's specific impulse, one with roots that go back to Project Daedalus. Pellets of deuterium and tritium would be imploded to produce power, using antimatter to light the reaction. The pellets themselves would be coated with uranium and then tungsten, while containing a small chip of uranium to one side of the hollow core. These would be dropped into a combustion chamber and struck by a stream of antiprotons, triggering fission in the uranium and lighting the deuterium/tritium fusion reaction.

As profligate with fuel as Daedalus, this engine would use quite a few pellets—the current estimate is 136 of the marble-sized pieces per second, allowing extended accelerations at one-fifth of Earth gravity. The inner planets, according to Brice Cassenti of United Technologies Research Center in East Hartford, Connecticut, would be within a week's voyage, while Jupiter would be only a month away. Mission times like that are just what a solar system-wide industrialization project would require. But while GDM uses antimatter to trigger the fusion reaction, it is in no sense an antimatter drive. The concept acknowledges profound limitations like our current inability to manufacture more than the tiniest amounts of antimatter. Rather than being the motive force of the engine, antimatter is just the spark that sets the machine in motion.

Producing no more than nanograms (one-billionth of a gram) of antimatter per year, we had better get used to working with small quantities, at least for the immediate future. But as the poet says, necessity is the mother of invention, and as is so often the case in interstellar studies, one provocative concept cross-fertilizes another. An interstellar propulsion system with great promise, for example, is the lightsail, a vast structure kilometers to the side pushed by laser or microwave energy from a power station in solar orbit (we'll examine lightsails in Chapter 6). But what if you could combine a lightsail with antimatter in such a way as to shrink the sail and do away with the need for an expensive, space-based power grid to push it? And what if you could build such a sail using amounts of antimatter that should be available within a decade? One last "what if": What if you could find a way to store antimatter that was relatively lightweight and easy to transport?

Coming up with answers to those questions would revolutionize propulsion, and the antimatter sail may do just that. In a sense, it is a refinement of the AIMStar antimatter/fusion engine, because it became clear as AIMStar was analyzed that the sheer weight of the needed containment magnets would be an impediment to building the thing. Steven D. Howe, who had worked on the AIMStar concept, began looking for ways to lose that weight. Howe, who has spent twenty years at Los Alamos National Laboratory, is a founder, along with AIMStar's Gerald Smith, of a company called Synergistic Technologies, where he has focused on antimatter containment. He is now CEO of a Chicago firm called Hbar Technologies. The name comes from physics notation, in which, the antiproton is written as a "p" with a bar over it ($\bar{p}$), while antihydrogen—a major focus of Howe's work—is an "H" with a bar over it ($\bar{H}$). Howe's new company focuses on the use of antiprotons for applications in everything from homeland defense to medicine. A science fiction novelist as well as physicist, Howe became fascinated along the way with advanced space propulsion.

Drawing Howe onto my radar was his recently completed Phase I study for NASA's Institute for Advanced Concepts on antimatter-driven propulsion systems for unmanned space probes, with an experiment-driven Phase II study now pending. That he manages to sandwich science fiction around both projects—as a contributor to *Analog Science Fiction and Fact* and author of a novel about commercializing space called *Honor Bound Honor Born*—also caught my eye. Many scientists seem to have been catalyzed by science fiction they read in their youth, and many have gone on to write it themselves.

Howe is a man who likes to take the long view. He believes that human history can be depicted in stages according to the kind of energy we burn. We have gone from fire to gunpowder to liquid combustion and are now edging into nuclear technologies that will one day open up the solar system. And one day, we'll move to antimatter.

Howe married the concept of antimatter to that of lightsails in his NIAC report "Antimatter Driven Sail for Deep Space Missions." His near-term target is to send a spacecraft some 250 AU from the Earth, out to the still-mysterious Kuiper Belt. Long term, he's thinking interstellar. And rather than relying on a laser beam pumped by a solar energy installation in orbit around the Sun, which is how most lightsail concepts are devised, Howe's sail is driven by antimatter released from the spacecraft itself. Its acceleration results not only from the collision of the antimatter with the surface of the sail, but also by nuclear fission, for Howe's sail is coated with a layer of uranium-235. The goal is to reach an average velocity of 116.8 kilometers per second (72 miles per second), which would deliver the 10-kilogram payload to the 250 AU target area in roughly ten years.

The antimatter-driven sail has some antecedents in Project Orion, which for a doomed project seems to have spawned more interest and subsequent theorizing than almost any deep space investigation. As a way of shedding the dead weight accumulated by the magnets used to confine antimatter in AIMStar, Howe and fellow scientists at Los Alamos began to consider antiproton-induced fission. One design, created by Johndale Solem and dubbed Medusa, featured nuclear weapons being set off behind a sail. The vessel would be pushed Orion-style, using the sail rather than giant pusher plates to catch the force of the explosions. But detonating a nuclear device between sail and spacecraft proved too daunting a challenge, particularly given the tethers and other apparatus needed to manage the sail effectively from the spacecraft.

Out of this came a much gentler method: Let antiprotons drift in the form of an antimatter gas across to the sail, hitting it at almost zero relative speed. The antiprotons meet a carbon-based sail that has been coated with uranium, thereby causing fission in the uranium. Some of the particles blown off the sail by fission are driven into space; the others are blown into the sail itself, imparting their momentum to drive it forward. The method depends upon storing antihydrogen—an antimatter atom consisting of an antiproton orbited by a positron—in the form of frozen pellets, or "snowflakes," which would evaporate on their way to the sail. As it comes into full operation, the sail material would heat to some 600 degrees Fahrenheit, and the antimatter cloud would be distributed across its diameter. Howe's ongoing Phase II study for NIAC involves experiments on uranium-laden foils, letting antiprotons hit them to gauge how much force the resulting reaction imparts. "We need," says Howe, "to measure the impulse delivered per antiproton per gram. That will allow us to then flesh out the entire architecture for both missions, the Oort Cloud and Alpha Centauri."

The antimatter sail would be no more than 15 feet in diameter, but the meeting of antihydrogen with sail provides plenty of punch. The trick is to create enough antimatter to make a viable fuel. Thus far, the amount of antimatter being produced is minimal, and its cost is huge—about $62.5 trillion per gram, or $1.75 quadrillion per ounce. According to Howe's NIAC report, the present global production of antiprotons is 84 nanograms per year.

So daunting is antimatter production that the cost of making antiprotons through particle beam collisions amounts to ten billion times more energy than is stored in their mass. CERN estimates that to create a kilogram of antimatter with present methods would take all the energy produced on the Earth for ten million years. Considered in more everyday terms, the amount of antimatter produced each year in an accelerator laboratory like CERN or Fermilab is about enough to make a 100-watt bulb shine for fifteen minutes. We're still a long way from a space drive.

But Howe points out that until about ten years ago, the production of antimatter was increasing by a factor of ten every two years, and there is reason to hope that it can be increased again. High-energy physics is all about particle collisions, such as those used in the study of quarks, among the smallest components of matter yet found. But if the goal were to isolate not just high-energy antiprotons—to reaccelerate them for more particle collisions—but to recirculate and capture all antiprotons available, then the amount of antimatter produced would increase by three or even four orders of magnitude. Howe believes that with money and motivation, scaling up antimatter production to the milligrams of antimatter needed for a Kuiper Belt mission would be feasible in the near term. An Alpha Centauri mission would require not just milligrams but grams of antimatter, a far more challenging task, but feasible within decades through upgrades to existing facilities and the creation of new, dedicated accelerators.

Storing the antimatter sail's fuel in pellet form makes it easier to manage than the alternative method, which involves holding the antihydrogen in a container of frozen hydrogen through the use of temperature and magnetic fields that prevent it from contacting the chamber walls and annihilating. For the amount of energy Howe envisions for his deep space mission, antimatter storage is itself a major risk—a microgram of antimatter contains billions of joules of power. And antimatter normally demands huge voltages to maintain the containment environment, a difficult feat to duplicate aboard a spacecraft.

But Howe's solution is elegant. He envisions storing the tiny pellets of antihydrogen in micro-traps that are built on integrated circuit chips using the kind of etching technology that produces today's microprocessors. Each chip would have a series of tunnels running the length of the chip, with small scallops, or dimples, creating wells at periodic intervals in which the antihydrogen pellets would be held. This storage system would be housed twelve meters away from the sail by four tethers. "Now you've got an integrated circuit chip that has a huge amount of energy," Howe said. "You can control it and thereby migrate the energy to where you need it. By changing the voltage on each scallop, you can move the pellets down peristaltically so that one pops out the end. Then you accelerate it up to speed and out, so that it drifts toward the sail."

At Marshall Space Flight Center, NASA's John Cole expressed to me more enthusiasm for the antimatter sail than any other propulsion concept we had discussed. At stake was the issue of energy, "specific energy," to be exact, which refers to the amount of energy available from fuel. As in 10 kilowatts per kilogram, a figure Cole says would allow manned missions to the outer planets with a two to three year round-trip time. "Howe's proposal puts a puff of antimatter out there and the antimatter initiates reaction with the U-238 in his sail," Cole said. "Now that releases a lot of neutrons and you get some secondary emissions. All those fission fragments leave the sail at enormous velocity, giving you a push that is essentially an ablation, but a nuclear-stimulated ablation. And the specific power this produces is something on order of 2,000 kilowatts per kilogram, which is the highest I've ever seen. It's just an enormous figure. So even if Howe's figures are an order of magnitude off, even two orders of magnitude off, a factor of 100, he is still in the realm of where we can have human exploration of the outer planets."

The pieces for the antimatter sail may be falling into place. In September of 2002, scientists at CERN's antiproton decelerator facility announced that they had created, for the first time, large numbers of antihydrogen atoms. The trick required mixing cold clouds of trapped positrons and antiprotons. Earlier experiments at both CERN and Fermilab had been able to create only a few energetic antihydrogen atoms per day, each moving close to the speed of light. The new antihydrogen experiment succeeded by slowing antiprotons, trapping them in an electromagnetic field, and mixing them with cold positrons. The result is cold—that is, very slow—antihydrogen atoms. In addition to their possible use as a spacecraft propellant, such atoms will allow close study of the differences between hydrogen and antihydrogen, perhaps leading to an understanding of why nature seems to prefer matter over antimatter.

The CERN team stored the antihydrogen it produced in Penning traps. To follow a roadmap of future experimentation toward the antimatter sail, what is now needed, in Howe's view, is for longer-term storage to be demonstrated in another kind of trap called an Ioffe-Pritchard trap. This would pave the way for the manufacture of antihydrogen pellets, the enabling technology for the radical storage concepts of the antimatter sail.

Someday antimatter may be abundant. If so, it will probably be because we have taken Robert Forward's suggestion to create actual antimatter factories, rather than relying on the production of antimatter in the particle accelerators of laboratories with much wider agendas. Forward is worth quoting on this point (particularly for his dig at scientists vs. engineers): "In a study I carried out for the Air Force Rocket Propulsion Laboratory, I showed that if an antiproton factory were designed properly by engineers, instead of by scientists with limited budgets and in a hurry to win a Nobel prize, the present energy efficiency (electrical energy in compared to antimatter annihilation energy out) could be raised from a part in sixty million to a part in ten thousand, or 0.01%, while at the same time, the cost of building the factory could be substantially lowered compared to the cost of the high precision scientific machines. From these studies, I estimated the cost of the antimatter at ten million dollars per milligram." If that still sounds like a lot, Forward points out that the energy efficiency of antimatter is such that a milligram of the stuff produces the same energy as twenty tons of chemical fuel. At $10 million per milligram, antimatter actually becomes more cost effective than chemical propulsion when used in orbit or deep space. Lifting twenty tons of fuel to Earth orbit via the Space Shuttle would cost roughly $100 million, compared to the $10 million price of our milligram of antimatter.

And if we have thus far been limited to considering antimatter designs that mix fission and fusion with the tiniest possible amounts of antimatter, the development of antimatter factories would widen the options considerably. It becomes possible to imagine a rocket that uses much larger amounts of antimatter. Various concepts are bruited about, beginning with a "solid core" drive that works by the annihilation of normal matter with antimatter inside a tungsten chamber. When heated by this reaction, the tungsten would in turn heat a hydrogen gas propellant that would be fed into the chamber. It is this hot gas, expelled from the rocket, that produces thrust. Variations on the same principle include a "gas core" drive, where the antimatter is annihilated directly in the propellant, with magnetic fields used to contain the charged pions from the reaction, and exhaust managed through a rocket nozzle. A "plasma core" drive would see antiprotons injected into a plasma (itself created by antimatter annihilation) that is contained by magnetic and electric fields. As the plasma was annihilated, it would heat a propellant for thrust. Note that in each case, we are relying upon the matter/antimatter annihilation to heat a propellant, with the plasma concept achieving the highest specific impulse because it would use only a magnetic nozzle for the exhaust, thus escaping the temperature limitations of a physical exhaust nozzle.

But the ultimate prize is a "beam-core" antimatter drive, where the rocket uses the pure product of the antimatter annihilation as its source of thrust. As protons and antiprotons meet, charged pions and gamma rays are produced. Because of their charge, the pions can be manipulated by magnetic fields that focus and funnel them down an exhaust nozzle. Conceivable specific impulse as high as ten million seconds (though with low thrust levels) would now be in range. Here we have the closest possible approach to Eugen Sänger's photon drive, a rocket whose exhaust velocity approaches that of light, made possible because we are not using the annihilation of electrons and positrons, which produces only gamma rays, but that of protons and antiprotons. The production of charged pions is what makes the difference, allowing the thrust to be channeled down the magnetic exhaust nozzle. A beam-core drive may never be built, because it depends upon the production of huge amounts of antimatter, but if it does become possible, it will allow robotic probes and even manned missions to neighboring star systems.

Nonetheless, even assuming abundant antimatter, a beam-core drive is hardly practical. The charged pions such an engine channels for its thrust must be controlled by magnetic fields before being ejected out the back of the spacecraft. To manipulate them properly requires an engine kilometers long. At a recent conference in Huntsville, the Jet Propulsion Laboratory's Robert Frisbee showed that such a spacecraft might have to be as much as 100 kilometers in length. While satisfyingly reminiscent of an old Cordwainer Smith story called "Golden the Ship Was—Oh! Oh! Oh!" in which a scarecrow of a spaceship millions of miles long seems to threaten the solar system, such a rocket pushes engineering to the breaking point.

But there is one propulsion system for the Alpha Centauri journey that does not rely on vast ships or materials we do not know how to produce. It is the work of Terry Kammash, whose work on magnetically insulated inertial confinement fusion (MICF) we considered in the last chapter. Kammash spoke of using antiprotons to catalyze a fusion reaction that might power an interstellar precursor mission, reaching the Oort Cloud within the lifetime of a researcher. That's a valuable step on the road to the stars, but in terms of specific impulse, it's not enough. Kammash points out that the best a fusion reaction can do is a specific impulse of about 100,000 seconds. That sounds like a lot compared to the Space Shuttle's 450 seconds, but it's far short of a Centauri mission. "For Alpha Centauri, you'll be going into the millions of seconds," Kammash said. "A pure antimatter engine could do that, but it would be kilometers long, not to mention the sheer amount of antimatter such a ship would require."

But Kammash has an alternative, one that uses lasers to accelerate charged particles to velocities close to the speed of light. Called LAPPS—Laser Accelerated Plasma Propulsion System—the engine works by focusing a powerful laser on a series of tiny targets, causing them to eject highly energetic particles. The intensity of the laser (measured as the power of the laser divided by the area of the target) is immense. "Even though the energy of the laser is just a few joules or a fraction of a joule, it's the power that counts," said Kammash. "We only use it for a femtosecond, or ten to the minus 15 seconds. You wind up with the power of the Hoover Dam with just a small laser dumped on a very small target." In other words, even a small amount of energy released in an extremely short period of time yields extraordinary power, power being defined as the rate by which energy is delivered. Concentrating rapid-fire lasers on micron-sized surfaces—a phenomenon known as "extreme light"—produces electric fields in the range of a trillion volts per centimeter, re-creating conditions found only at the center of stars.

Lasers like this are the latest word in research for everything from astrophysics to medical radiography, and play a key role in the inertial confinement fusion concepts we discussed in Chapter 3. In the LAPPS scheme, the pulsed laser is powered by electrical energy created by an onboard nuclear reactor. The target: minute aluminum strips coated with fluorine or carbon. As the strips are fed a thousand times a second into the combustion chamber, the synchronized laser pulses strike them, kicking loose streams of protons that are the source of the spacecraft's thrust. The pulses of light push the charged particles to relativistic speeds almost instantly, accelerating them at 10,000 times the rate of today's particle accelerators. Thus we segue out of antimatter, though drawing inspiration from systems like inertial confinement fusion that use it. Research into one propulsion concept feeds the others. Kammash recently worked out the kinks in his proposal in a study for NASA's Institute for Advanced Concepts.

A fully functional LAPPS engine would eject protons at a high percentage of the speed of light, while generating a specific impulse of between three and four million seconds. Laboratory work has proven to Kammash that the principle is sound. His next project for NIAC is a Phase II study designed to find ways to increase the thrust of the system. There are two ways to do that: by ejecting heavier particles, or by increasing the number of particles ejected by the laser. Right now LAPPS measures thrust in tiny fractions of a newton (a newton is the unit of force needed to impart an acceleration of one meter per second per second to a mass of one kilogram). But Kammash says that increasing that thrust to 100 newtons, the equivalent of about 22 pounds of thrust, would be sufficient to send a robotic probe to Alpha Centauri for a rendezvous mission in 100 years.

"For an Alpha Centauri mission, this approach is based on present day understanding," Kammash told me. "We understand the physics. What we have to do is improve the technology to be able to produce that kind of thrust. And that is where we stand. Can we produce high enough thrust by bombarding bigger and bigger targets? And if we succeed in doing that, then we'll have a system that could make it in a hundred years instead of a hundred thousand years."

Kammash, who also worked on the gas dynamic mirror fusion experiments at Marshall Space Flight Center, continues to push the boundaries of what will be possible within the next half-decade. As we'll see later, this is precisely what NASA's Institute for Advanced Concepts was set up to do, and what scores of scientists continue to pursue, some experimentally at government research centers, others in private businesses and universities, often on their own time. An outsider can be forgiven the notion that researchers are throwing ideas at the interstellar problem just to see what will stick, in the hope of finding a breakthrough that makes all previous work obsolete. But discoveries get made that way. And if there could be said to be a human interstellar mission, it is at present a collective, theoretical effort not united by common funding as much as energized by a common vision, one that is all about reaching distant places with missions that seem impossible. Whether or not we achieve interstellar travel, we still seem to be hardwired to need the imagining of it.

CHAPTER 5

JOURNEY BY STARLIGHT: THE STORY OF THE SOLAR SAIL

The tools of interstellar flight are, one by one, being defined. They are hardly ready to fly over such distances, but even in the case of propulsion, the most intractable problem of all, solutions are being suggested and ideas tested out in laboratories. If our goal is to build a robotic interstellar probe that can deliver a payload to the nearest star and return data from it, then the evidence of our first fifty years of space exploration is that a 4.3-light-year journey is no longer out of the question. The solar sail technology that will probably be our first interstellar propulsion system is one of breathtaking elegance. It will take us to the stars on a beam of light.

I readily remember the day I first encountered this technology. The date was December 7, 1971, and the temperature was an utterly unseasonable 70 degrees in Grinnell, Iowa. It was one of those days so wildly out of sync with normal weather patterns that you had the feeling disaster loomed, and sure enough, by late evening the temperature had plummeted well below freezing and a massive ice storm descended upon the town. A college senior, I was living with my wife in a drafty house on High Street, and at 11 P.M. trying to stay warm under a quilt while I read by the light of a small bedside lamp. Outside, pellets of ice were striking the eaves, but later the sound would become the thumps of falling branches and power lines.

For the moment, though, we still had power, and the sound of the falling ice pinging against the shingles seemed as if it came from another universe. I had just discovered spaceflight by sail in an extraordinary story called "The Lady Who Sailed the Soul." Written by Cordwainer Smith, the story appeared in an eleven-year-old back issue of *Galaxy Science Fiction* I had found in a local bookstore. It told the tale of a space sailor named Helen America, who had fallen in love with the strange Mr. Grey-no-more, lost him when he returned to the stars, and tracked him down after a space voyage of some forty years.

Her ship was called the *Soul.* Smith wrote of "... the image of the great sails, tissue-metal wings with which the bodies of people finally fluttered out among the stars." A later Smith story, "Think Blue, Count Two," also published by H. L. Gold in *Galaxy* (February 1963), laid out the age of lightsails: "Before the great ships whispered between the stars by means of planoforming, people had to fly from star to star with immense sails—huge films assorted in space on long, rigid, coldproof rigging. A small spaceboat provided room for a sailor to handle the sails, check the course and watch the passengers who were sealed, like knots in immense threads, in their little adiabatic pods which trailed behind the ship. The passengers knew nothing, except for going to sleep on Earth and waking up on a strange new world forty, fifty, or two hundred years later.... This was a primitive way to do it, but it worked."

Scholar, diplomat, soldier, and occasional spy, Cordwainer Smith was a *nom de plume* for Paul Myron Anthony Linebarger. A man who wore a fedora and tinted glasses (he was plagued with vision problems throughout his life), Linebarger grew up in Washington, D.C., Baden Baden (Germany), and Shanghai; his father was an advisor to Sun Yat-sen. Among other events in a crowded life, he served in the U.S. Army in China and India, and taught Asiatic studies at Johns Hopkins, Duke, and the School of Advanced International Studies in Washington, where he worked with the CIA on the side. At one point, in 1952, he worked with Howard Hunt (of Watergate notoriety) in Mexico City, and developed propaganda techniques and psychological warfare methods for U.S. intelligence. Somehow he found time, in addition to books on Far Eastern affairs, to write actively for science fiction magazines between 1950 and 1966, when he died. A committed high Episcopalian, Linebarger worked religious themes and allegories into much of his fiction.

How this polymath came across the notion of solar sails may never be known, but as an active science fiction reader, he may well have run across a seminal article on the technology that appeared in the May 1951 issue of *Astounding Science Fiction*, in which an engineer named Carl Wiley laid out the basics of propulsion by photons. Whatever the case, the elegance of his language (sometimes using Chinese narrative techniques) and his breathtaking imagery have fired the imagination of several generations of readers. And if Linebarger's probing and allusive work was never long on science, that would soon be remedied by other science fiction writers, some of them scientists themselves. It was probably Arthur C. Clarke more than anyone else who put space sails into the public consciousness, but it was Smith who put them into our psyche. That night in Iowa I dreamed of winds off the prairie filling gossamer sails larger than continents.

It would not be long before such dreams, using not wind but the pressure of photons, began to emerge in the pages of prominent astronautical journals. From *Acta Astronautica* to *The Journal of Spacecraft and Rockets* and, especially, the *Journal of the British Interplanetary Society*, their pages fairly crackled with excitement over the sail concept. It was here in the 1980s that physicist Robert Forward began to develop the idea that interstellar travel was truly feasible, that using physics understood today, we could engineer and build huge structures, sails hundreds of kilometers in diameter, and propel them by a laser beam or perhaps a microwave beam to speeds high enough to reach the nearest stars. The thinking was bold, to say the least: Imagine a space-borne lens the size of Texas that focuses the light from thousands of solar collectors. But the physics worked. Forward was now saying that one-tenth of the speed of light was within reach. A craft flying at that velocity reaches Alpha Centauri in 43 years. Add another 4.3 years for data return and you have a mission whose outcome can be witnessed by many of those who launched it.

The idea danced around and through science fiction, a field from which Forward, who wrote science fiction himself, obviously drew inspiration. Indeed, he was given to publishing nonfiction articles in the pages of *Analog Science Fiction and Fact*, outlining the state of the art on novel propulsion methods or new ways of looking at gravity. He worked shortly after Linebarger's era, surely captivated by the 5,000-kilometer sails of "Think Blue, Count Two," and was an inspiration for the laser-driven sails of author Larry Niven, who used them in his *Tales of Known Space* series. One Niven novel, *The Mote in God's Eye*, written in collaboration with Jerry Pournelle, posits man's first encounter with an alien species upon the arrival of a huge lightsail, the "mote" being the laser beam pushing the ship from a world in the Coal Sack Nebula as seen by human observers on the distant colony world of New Scotland. Niven and Pournelle derived the idea of laser-pushed lightsails from their conversations with Forward.

Forward, who worked at Hughes Research Laboratories, where the laser was invented, would study the interstellar sail concept for four decades, spinning off ideas as exotic as Starwisp, a spider web of a robotic probe made of a mesh one kilometer in diameter and weighing only sixteen grams, with embedded circuitry as its payload. And in his novels, especially *Rocheworld*, he would show that sails could be constructed that allowed a spacecraft to be decelerated at a distant solar system rather than just flown headlong through it. Countless other scientists rose to the interstellar challenge, motivated by his example.

Forward knew that the principle of solar sailing was anything but science fiction, having already been demonstrated in space. Consider the case of Mariner 10. Launched by an Atlas/Centaur rocket in late 1973, two years after that Iowa ice storm drove "The Lady Who Sailed the Soul" across my personal cosmos, the doughty NASA spacecraft made three passes near the planet Mercury, using a gravity assist around Venus to bend its orbit inward toward the Sun's closest planet. Mariner 10 carried a host of scientific instruments, including cameras, an ultraviolet spectrometer, an infrared radiometer, and various tools for measuring the solar plasma and magnetic fields. The spacecraft encountered Mercury for the first time at less than 500 miles from the planet's scorched surface. It then entered a solar orbit that permitted two more passes, with the third pass the closest of all, at just over 200 miles. From these vantage points, Mariner 10 was able to photograph about half of this barren, moon-like world.

A photograph of Mariner 10 is dominated by the two solar panels that jut out from each side of the spacecraft. Each panel measures close to 9 feet in length and slightly over 3 feet in width, creating a usable area of 55 square feet for its power-generating solar cells. Spacecraft don't have to be pretty, and the Mariner series would never be described as such, but the sheer functionality of the design had already shown its stuff, with successful flybys of Mars starting with Mariner 4 in 1965, and a triumphant Mars orbital mission in 1971, making Mariner 9 the first spacecraft to orbit another planet.

Mariner 10's spectacular night launch from Cape Canaveral gave little clue of the trouble that was to follow, or the unexpected means by which its problems would be resolved. The high-gain antenna malfunctioned early, later mysteriously coming back to life in the warmer environment of Mercury. Stabilizing gyroscopes proved balky and unpredictable, and the spacecraft was operating on backup power before it even reached Venus. An even greater problem was flaking paint from Mariner 10's high-gain antenna, which confused its navigation sensors. The spacecraft used the star Canopus for celestial guidance, but bright particles of paint caused it to lose its fix on the star. More serious trouble showed up on March 6, when the disturbed star tracker caused the spacecraft to roll. For forty long minutes, the onboard gyroscopes were activated and precious attitude control gas was vented into space.

But Mariner's guidance and control team at Pasadena's Jet Propulsion Laboratory had a fix in place by March 11. In the dry language of the mission report:

> On 11 March the spacecraft was not oriented in roll on the star Canopus as it went out of the star sensor's view early in the morning due to the distraction of stray bright particles. However, the spacecraft was stable in its roll attitude because the tilt of Mariner's solar panels is such that solar pressure on the panels created a torque counter to the natural drift in roll of the spacecraft. This stable condition was the result of skilled testing over the weekend to develop various combinations of panel tilt angles. One panel is at 66 deg and the other at 66½ degrees. The Mariner 10 roll attitude can be controlled by judicious tilting of the spacecraft solar panels.

What you do when spacecraft lose systems is try to figure out what still works. Fortunately for planetary science, those big solar arrays were operational. Their photovoltaic cells were built to absorb sunlight and create a current flow between two layers of opposite charge. Usually built of crystalline silicon and gallium arsenide, solar panels get the most power when pointed directly at the Sun, and are therefore installed in such a way that the spacecraft can pivot them to keep their solar orientation fixed. The Mariner team had figured out how to rotate the craft's panels to take advantage of the photons flooding them from the Sun, using the pressure of light to make this mission the first practical demonstration of solar sailing.

The Mariner 10 maneuver would surely have won the approval of Johannes Kepler (1571–1630), had this strange, obsessive man been around to see it. Shaped in childhood by his viewing of the Great Comet of 1577, Kepler went on to use Tycho Brahe's observational data to deduce the proper shape of planetary orbits: the ellipse. A deeply religious thinker who looked for divine archetypes in the natural world, Kepler was convinced that the universe moved through a mathematical plan, and that numbers could unlock the secrets of the cosmos. Ultimately, three laws of planetary motion would be attached to his name, and so would the first inkling that sunlight can carry a propulsive force.

If Copernicus had established that the sun was at the center of the heavens, it was Kepler who set the planetary system in motion. His *Epitome Astronomiae Copernicanae* (1618–21) was the first genuine textbook of Copernican astronomy, incorporating his laws of planetary motion, which he regarded as observations of celestial harmonies. Remarkably, this theorist, imperial mathematician, and occasional astrologer was also a science fiction writer. In fact, in an address to the British Interplanetary Society, fellow writer and scientist Arthur C. Clarke acknowledged Kepler as one of his literary forebears, placing him in a long tradition that culminated in modern science fiction.

Published after Kepler's death in 1634, the *Somnium* ("Dream") is an account of life on the Moon. It is a story stuffed with ideas, and has been considered the first example of so-called "hard" science fiction, a genre dependent upon scientific theory and calculations. The *Somnium* tells the story of a young Icelander named Duracotus, a student of Tycho Brahe, who travels to the moon aided by the witch-like powers of his mother. It is interesting to speculate on what resonance his own mother's powers had with Kepler, since she was later tried for, though not convicted of, witchcraft on the evidence of the *Somnium*. (Kepler's work provided the ideal excuse for the enemies this opinionated, difficult woman had collected over the years.) The tale itself invokes the shock of launch, problems of weightlessness and life support, lunar geography, and extraterrestrial beings: the Moon turns out to be inhabited by a race of serpent-like creatures. Kepler added a series of notes that further examined the mathematical and philosophical ideas in the work (his notes, in fact, are four times longer than his text) and an appendix stuffed with astronomical calculations. It was a strategy that would win favor with readers of twentieth-century science fiction magazine editors like John Campbell, whose *Astounding Science Fiction* regularly interleaved essays by scientists among its stories and serialized novels.

Kepler the Author's use of witchcraft to get Duracotus to the Moon was a plot device to advance his story, but Kepler the Scientist also had a deep interest in alternate forms of cosmic voyaging. A key player in the development of the science of optics, he remembered the Great Comet of 1577, today preserved in a spectacular engraving by Jiri Daschitzky, showing a fiery storm filling the sky over Prague on November 12 of that year. Kepler would have been six years old when he saw the comet from the German village of Weil der Stadt, his hometown. Memories of that object must have been reawakened by the appearance of another comet in 1618, one bright enough to cause gold and silver ducats with comet motifs to be minted in Frankfurt late in that year.

Later, Kepler would study such comets, noting that their tails seemed to react to what appeared to be a solar breeze. We know now that the sublimation of water ice and embedded dust from the comet's surface, upon heating as the comet approaches the Sun, creates an atmosphere around the core that feeds into the tail. Kepler noted that the tail of a comet always pointed away from the Sun, whether the comet was approaching it or leaving.

Was there some kind of pressure analogous to a wind blowing through the celestial regions? If so, it might be possible one day to harness that wind to travel through space. Kepler pondered the issue in a letter to Galileo written in 1610: "There will certainly be no lack of human pioneers when we have mastered the art of flight.... Let us create vessels and sails adjusted to the heavenly ether, and there will be plenty of people unafraid of the empty wastes. In the meantime, we shall prepare for the brave sky-travelers maps of the celestial bodies. I shall do it for the moon, you Galileo, for Jupiter." Kepler went on to explore optical and cometary phenomena in two books, *De Cometis libelli tres* (1619) and *Hyperaspistes* (1624).

René Descartes would suggest within decades that light was indeed a form of pressure, seeing it as a wave moving through an invisible medium called ether. By 1903, with the ether concept about to be definitively shattered by Einstein, the Swedish chemist and Nobel prize winner Svante Arrhenius invoked a different pressure mechanism, that of solar light. Arrhenius argued that life arrived on Earth in the form of microscopic spores that had been pushed across the interstellar void by the pressure of starlight.

Science fiction would explore light as a propulsive force repeatedly in the latter half of the twentieth century, as we'll see, but the first science fiction story that made the physics of light central to its theme was an obscure novel called *Aventures extraordinaires d'un savant russe* (*The Extraordinary Adventures of a Russian Scientist*), by Georges Le Faure and Henri de Graffigny. Published in three volumes (1889–91) and illustrated by astronomer Camille Flammarion, who also wrote a preface to each volume, *Aventures* drew on Jules Verne's *De la terre à la lune* (*From the Earth to the Moon*, 1865) and *Autour de la lune* (*Around the Moon*, 1870). A fourth volume by Le Faure and de Graffigny was subsequently released under the title *Les mondes stellaires* (*Stellar Worlds*).

Both Le Faure and de Graffigny had experimented with rocketry, and used a method familiar to Verne in their novel, that of launching their moon ship from a cannon set into a volcano, with lava as the propulsive agent. But it is their subsequent voyaging that interests us. Venturing from the Moon to Venus and on to Mercury, our heroes travel in a hollow sphere pushed by the pressure of light from the Sun. Launching the sphere is accomplished by concentrating sunlight from a huge reflecting dish onto a selenium disk, creating a final speed of 28,000 kilometers per hour, a speed that compares not unfavorably with that of Voyager 1, now pushing out toward the Kuiper Belt at some 62,000 kilometers per hour.

Remarkably, the final volume took the space travelers to Jupiter in a vessel that sucked up interplanetary particles and pushed them out the back, employing a scheme much later theorized by Robert W. Bussard. Does the story inspire the theorist, or, more likely, simply set a pool of thought into agitation, so that its widening rings become visible from all nearby shores? It is unlikely, though by no means impossible, that Konstantin Tsiolkovsky read Le Faure and de Graffigny, but it was the father of Russian space science who gave scientific weight to the idea that light could propel a spacecraft, supported by his Latvian colleague Friderikh Arturovich Tsander. Tsander's work, describing huge mirrors made of very thin materials that used photons from the Sun to achieve high speeds, appeared in the early 1920s, though it was not widely circulated.

While James Clerk Maxwell had predicted in 1864 that light exerted pressure, based on his theory of electromagnetic fields, it was a Russian physicist named Peter Lebedev who first demonstrated that pressure in the laboratory. Lebedev experimented with metal sheets of differing levels of reflectivity, some black, some bright as mirrors, mounted in a vacuum chamber. Measuring the torque the various surfaces put on a supporting rod when strong beams of light were focused upon them, he was able to show values for light pressure that came close to Maxwell's predictions, thus supporting the electromagnetic theory of light. Along the way, by showing that light could exert forces upon objects in a vacuum, he corroborated Kepler's thesis about the displacement of comet's tails. Albert Einstein would go on to win the Nobel Prize in physics in 1921 for his 1905 paper on the photoelectric effect, which, among other things, explained how light exerts such a force.

It would take a popular book to turn the attention of more than a few specialists to the idea of solar sailing. In 1929, British physicist, crystallographer, and fervent leftist John D. Bernal published *The World, the Flesh & the Devil: An Enquiry into the Future of the Three Enemies of the Rational Soul*. There he opined that light might become the operative force for future propulsion:

> However it is effected, the first leaving of the earth will have provided us with the means of traveling through space with considerable acceleration and, therefore, the possibility of obtaining great velocities—even if the acceleration can only be maintained for a short time. If the problem of the utilization of solar energy has by that time been solved, the movement of these space vessels can be maintained indefinitely. Failing this, a form of space sailing might be developed which used the repulsive effect of the sun's rays instead of wind. A space vessel spreading its large, metallic wings, acres in extent, to the full, might be blown to the limit of Neptune's orbit. Then, to increase its speed, it would tack, close-hauled, down the gravitational field, spreading full sail again as it rushed past the sun.

No more prescient account of what physicist/novelist Gregory Benford has called a "sun-diver" maneuver has ever been written: The craft closes to less than a single solar radius, then slingshots around the Sun with a massive gravity assist, spreading its sail at the closest approach. Amazingly, Bernal wrote this at a time when the world's fastest vehicle was a black and red monoplane called the *Travel Air Mystery Ship*, which had achieved 235 miles per hour at the National Air Races in Cleveland!

Solar sails caught John Campbell's eye at *Astounding* in 1951, when an engineer named Carl Wiley wrote "Clipper Ships of Space," another of Campbell's nonfiction pieces designed to seed ideas among future authors. Possibly worried that his reputation might suffer from appearing in a science fiction magazine, Wiley wrote under the name Russell Saunders, discussing methods of steering and tacking such a craft. He went on to create a design for a solar sail some 80 kilometers in diameter that resembled a parachute. The concept was audacious for its time, but the 80-kilometer sail would soon be dwarfed by even larger light-gathering surfaces imagined by science fiction writers and scientists. By 1958, IBM's Richard Garwin was able to write a scientific paper in the journal *Jet Propulsion* that laid down the basic principles of solar sailing, the first essay on the subject to appear in the English language in a technical publication, and the beginning of modern work on the sail concept.

What will propel our first interstellar probe remains problematic, but variations on sail technology must certainly head the list. Gregory Matloff, who has led modern research into the sun-diver maneuver, realized in the early 1980's that it could be used to fly an interstellar mission in time frames that were, for the first time, within the scope of human history. "What I found out is what everybody knows now," Matloff told me when we spoke at Marshall Space Flight Center, where he was on a consulting assignment. "It is quite possible to take both large ships and small probes to the nearest star within a thousand years, using the sail alone. But that's about it for an unassisted sail. It is very difficult to get the mission time down below 800 or 900 years." Matloff's early paper on solar sails for interstellar missions, written with Eugene Mallove and called "Solar Sail Starships: The Clipper Ships of the Galaxy," is one of the core articles in the field of interstellar studies. Its title gives a generous nod to Wiley's early essay.

Solar sailing works because while photons have no mass, they do carry momentum. The force they exert is so slight that a sail one-third of a mile square in Earth orbit would be pushed to no more than ten miles per hour in its first hour of operation. At Earth's distance from the sun, the solar flux is on the order of 1.4 kilowatts per square meter, about enough to run a hair dryer, and nine orders of magnitude weaker than the force of the wind on the Earth's surface. The beauty of sunlight, of course, is that it's plentiful as long as you remain within a reasonable distance of the Sun, and it doesn't stop producing its effects within that environment. Each tiny push mounts up. Make the sail lightweight, free of distorting wrinkles, and large—some proposals for interstellar precursor missions call for sails half a kilometer across, while interstellar mission concepts can range into the hundreds of kilometers—and you've produced a surface that can accelerate and navigate by altering the position of the sail in relation to the Sun. The craft can be steered in a way vaguely similar to a sailboat, but one without a conventional rudder. Instead, it is the orbital velocity of the spacecraft in conjunction with the pressure of photons that allows the craft to be controlled.

The sailboat analogy is an imprecise one, and its use in popular fiction has caused some misunderstandings. A case in point is Arthur C. Clarke's "The Wind from the Sun," the story that, more than any other, put solar sailing into the public consciousness when it ran as a cover story in *Boy's Life* magazine back in March of 1964. Originally titled "Sunjammer," a wonderful nautical neologism (and one chosen, almost at the same time, by Poul Anderson for his own tale of solar sailing), "The Wind from the Sun" told the story of a race to the moon from Earth orbit involving spacecraft, with yachtsmanship as the key metaphor. Mixed with reports in the popular press of a so-called "solar wind," these sea-going references gave credence to the notion that it was a stream of solar particles rather than light itself that drove solar sails.

The solar wind is real; as we will soon see, it is the driver in at least one breakthrough propulsion design, the magsail (also known as M2P2, for Mini-Magnetospheric Plasma Propulsion) advocated by the University of Washington's Robert Winglee. But in terms of interacting with a large, reflective sail in the inner solar system, this stream of charged particles moving at speeds up to 500 kilometers per second is dwarfed by the cumulative force of photons transferring their momentum to the sail surface. Magsail concepts have their own virtues and these are considerable, but the kind of solar sailing Clarke was writing about doesn't use the solar wind for its push.

In "The Wind from the Sun," Clarke's sailor, John Merton, commands the vessel Diana, pulled by a sail fifty million square feet in size, linked to its command capsule by over a hundred miles of cabling. "All the canvas of all the tea clippers that had once raced like clouds across the China seas, sewn into one gigantic sheet, could not match the single sail that Diana had spread beneath the Sun," Clarke writes. "Yet it was little more substantial than a soap bubble; that two square miles of aluminized plastic was only a few millionths of an inch thick." The enormous sail catches sunlight hour after hour, day after day. In its first second under thrust, it moves about a fifth of an inch. After a minute, it has been pushed sixty feet. An hour later, the sail has moved forty miles, the spacecraft topping eighty miles per hour. In days, Diana has reached escape velocity and is on her way to the Moon.

Solar yachts on a race from the Earth to the Moon present a tantalizing scenario. The race sounds like it would make a great movie. But in Nazelles-Negron, east of Tours in the beautiful rolling terrain of France's Loire Valley, a group of dedicated solar sailors actually tried to promote such a race. Known as the Union pour la Promotion de la Propulsion Photonique (U3P), the group has been active since 1981, initially hawking a race much like Clarke's, in which effort it was joined by the Solar Sail Union of Japan. Official rules for the Luna Cup were approved by the International Astronautical Federation at the World Space Congress in August of 1992. The U3P doubtless drew inspiration from the work of French science fiction author Pierre Boulle, who wrote of solar sailing in his book *Planet of the Apes*. "Jinn and Phyllis, a wealthy and free couple, were known through the Cosmos to be young originals, with a bit of craziness, and they would cruise through the Universe just for the fun of it—with their sailcraft," Boulle wrote. "Their ship was a kind of sphere with a shell—the sail—made of amazingly thin material, and it would move through space, just pushed by the pressure of light beams."

The Moon race, and a later flurry of interest in a sail race to Mars, produced interesting designs even if none of them is being actively pursued. But the idea of using a race to drive technology development has a long and proud history in aeronautics. Louis Bleriot won the Daily Mail prize for being the first to cross the English Channel, while the Baron de Forrest prize offered £4,000 for the longest flight from England to the Continent; Tommy Sopwith won it in 1910 for his three-hour flight to Belgium. If the $25,000 Orteig Prize could motivate Charles Lindbergh to cross the Atlantic, surely a cost-adjusted prize could spur private industry in space research. And indeed, the Ansari X Prize, a $10 million award for the first team to build and fly a three-person vehicle to an altitude of 100 kilometers twice within a two-week period, is being eagerly contested. Twenty-five groups from seven countries are in the hunt. In 2004, one of them, Scaled Composites of Mojave, California, became the first to reach space with a manned, privately built craft, the SpaceShipOne rocket plane. Chances are strong that the prize will have been claimed by the time you read this.

Even without ongoing race preparations, U3P has been an active player in solar sail cooperation, working with such groups as Russia's Space Regatta Consortium and the Aero-Club de France. Indeed, interest in solar sailing is flourishing throughout Europe. The European Space Agency (ESA) is itself involved in a solar sail project, exploring technology development with the DLR German Aerospace Centre in Cologne. The agencies have settled upon a square sail design with diagonal booms supporting four triangular sail elements, the booms made of reinforced carbon fiber plastic. Two proposed orbital experiments using these designs have been considered. The first, called Daedalus, would remain in Earth orbit and demonstrate photon propulsion. The second, flying under the name ODISSEE (Orbital Demonstration of an Innovative Solar Sail-Driven Expandable-Structure Experiment), would make the Earth to Moon crossing, carrying high-resolution cameras in its scientific payload. DLR and ESA are developing (and have deployed on the ground) a 20 × 20-meter model of the spacecraft to demonstrate the feasibility of their design.

The fact that solar sails require no fuel aboard the spacecraft is perhaps the major reason for the intense interest in these craft on both sides of the Atlantic. Mention that factor to NASA's Les Johnson and he responds with an emphatic "Amen!" Johnson is manager of the In-Space Propulsion Program at NASA's Marshall Space Flight Center in Huntsville, Alabama. He used to head the even more provocatively titled Interstellar Propulsion Project, which has been folded into the In-Space effort. Marshall manages NASA's program to develop a solar sail, a project that encompasses several NASA centers including the Jet Propulsion

Laboratory and Langley Research Center in Hampton, Virginia, near the mouth of the Chesapeake Bay.

A straightforward, enthusiastic man, a science fiction reader from childhood, Johnson is trying to find ways to make propulsion systems better and more efficient. The solar sail is high on his list of candidates. "If you don't have to carry all that fuel with you, not only can you add science payload, but your launch vehicle can be smaller because it doesn't have as much weight to throw," Johnson told me. The name of the game at Marshall is what Johnson calls "primary propulsion," which means something that generates a big delta v, or change of velocity. Change of velocity doesn't have to equate to high thrust, because what counts on long voyages is steady, efficient pushing, a scenario the solar sail fits admirably. In addition to the solar sail, Johnson also focuses on electric propulsion systems and aero-assist methods that use planetary atmospheres to adjust orbits. The Mars Climate Orbiter used aero-assist methods upon its arrival at Mars in 1999 to change its highly elliptical orbit to a circular one, working with the drag exerted by the upper atmosphere on its solar array.

NASA operates through a series of technology readiness levels (TRLs), in ascending order from 1 to 9. TRL 1, says Johnson, is akin to being in the cafeteria with a Nobel laureate who scribbles an equation on his napkin. The physics seems sound and his fellow diners are enthusiastic, but the idea is still in the area of pure research. Marshall picks up the solar sail work at TRL 3, where the fundamentals of the physics have been demonstrated but the engineering has yet to be designed. Some solar sail elements are at TRL 3, Johnson adds, while others are higher, but in any case, the Marshall effort ends at TRL 6, the point at which the program can demonstrate a deployable sail that can be used for ground-based simulations and experiments, especially on the craft's guidance and control software—"everything you can do on the ground in a systems-level relevant environment," is how Johnson puts it. Whether NASA moves higher up the TRL scale to flight validation (TRL 9, at which the spacecraft is ready to fly) is a decision that is out of Marshall's hands.

NASA's experience with solar sails is surprisingly deep. In fact, it was the publication of Richard Garwin's 1958 paper in the journal *Jet Propulsion* that inspired early NASA technical papers. At a 1960 meeting at Langley Research Center, the agency examined sail designs, developing an early preference for spinning sails and, later, an interest in heliogyro designs, in which enormous blades of sail material rotate around a common axis. Even so, little practical work was being done by the 1970s except at Battelle Memorial Institute in Columbus, Ohio, where Jerome Wright was analyzing mission possibilities for the new technology. One mission stood out: the possibility of a rendezvous with Halley's Comet, due to be within range of Earth-based spacecraft some time in 1986. For this mission, the heliogyro design seemed preferable to a square sail because of the difficulties of deploying the latter. Deploying the heliogyro would itself have been a remarkable feat: The system would use twelve 7.5-kilometer blades of sail film that would have been unrolled individually.

Studies at the Jet Propulsion Laboratory in the mid-1970s revealed that the spacecraft for the comet rendezvous would have to be launched in 1981. Louis Friedman, later head of The Planetary Society (which he co-founded in 1980 with then-JPL director Bruce Murray and Carl Sagan) developed a proposal that led to intensive JPL work on sail design. What made such a mission theoretically possible was the assumption that the Space Shuttle would be operational by launch date, solving the problem of getting the large payload of the sail into orbit. But NASA administrators worried about the ticking clock that would have forced them to use an untested technology on a near-term mission, however spectacular. A solar-electric propulsion system was eventually chosen, but by that time the cometary mission itself was in jeopardy. Halley's Comet would finally be examined by spacecraft from the USSR, Japan, and the European Community, but no American craft joined them.

Even Louis Friedman, a solar sail advocate ever since, now believes the technology was not ready for that rendezvous, and subsequent events bear him out. Flying a sail is complicated business, as was demonstrated by the first sail deployment in outer space. This occurred on Russia's Znamya mission, with deployment on February 4, 1993. Designed as an experiment in beamed solar power, Znamya was created by the Space Regatta Consortium under sponsorship by the Rocket Space Corporation Energia, licensed by the Russian Space Agency. The Znamya was to demonstrate the practicality of beaming solar energy to polar and subarctic settlements. The 20-meter spinning sail-mirror was successfully deployed from a Progress supply ship that had just undocked from the Mir space station.

"Even so, Znamya didn't work very well," said Hoppy Price in his JPL office, showing me a photograph of the deployed sail. The object looked like nothing so much as a sand dollar washed up on a beach. "You can see there are scallops that weren't supposed to be there. And there are these wrinkles in the sail, so it didn't really work quite the way it was supposed to work, and it was a lot heavier than what we'd like to build. But the more we study this, the basic configuration—for a large spin-stabilized sail that would be very low in mass—is a promising one."

Problems like these are why ground experiments in sail deployment are crucial. At NASA, the In-Space Propulsion Program is investing approximately $10 million per year on solar sail research, hoping to develop a first-generation sail for interplanetary work. That sail may be ground tested as early as 2005, with the hope of getting it approved for a flight validation mission to test it in space. Just as NASA's Deep Space 1 mission spacecraft tested and validated its ion propulsion system in 1998, sail designers hope to see their craft flown and then approved for subsequent science missions like the Solar Polar Imager. Two contractors—L'Garde, Inc., a specialist in inflatable space structures in Tustin, California, and Able Engineering in nearby Goleta—are developing their own sail designs, either or both of which may be chosen for such missions.

At JPL itself, the delicate business of sail deployment is under active investigation. A solar sail mission to the outer planets would require a huge sail, and some of Robert Forward's concepts call for laser-beamed sails hundreds of kilometers in diameter. The trick is to pack as much material as possible into the smallest space aboard a launch vehicle, and having done that, to find a way to unfold the sail with unerring precision so as to avoid what happened to Znamya.

Each sail panel, flexible and microns-thin, must be unfurled without becoming entangled with the others. This is no small feat: a micron is a millionth of a meter; even a human hair is around

100 microns in diameter. Moreover, testing any sail configuration is a challenge since the tests have to be conducted on the ground, where both gravity and the presence of an atmosphere complicate the picture compared to a more straightforward space deployment. The best method for ground tests is to work with a small sail that can be deployed in a vacuum chamber.

JPL's Moktar Salama is a specialist in structural dynamics, the study of moving objects in relation to the physics that affect them. Salama has been working with a sail one meter in diameter, spinning it at approximately 200 rpm to overcome the effects of gravity. By using scaling laws, Salama can determine how his results would compare to a larger sail operating in the gravity-free vacuum of space. But the scientist told me that those laws must be applied with caution. "Our scaling laws account for what we know. But they do not account for things we don't know. Some scales introduce effects that don't exist. When the size is large, the scale effect is minimal. When it is small, it overwhelms the specimen. So there are scale effects we cannot violate, and we have to be circumspect."

Salama's work, in other words, is a fine balance, offsetting gravity with spin and sail size, while trying to produce a result not distorted by the dimensions of the laboratory sail. Working with 2.5-micron Mylar, a shiny, tough polyester film, the scientist was able to deploy the sail without entanglement. The next step will be to test deployment techniques in a weightless environment, which Salama hopes to do aboard a KC-135 research aircraft that can create twenty to twenty-five seconds of weightlessness by flying parabolic arcs. Typically used for refueling military aircraft, the KC-135 has been modified by NASA to conduct microgravity research, and to train astronauts (who have famously dubbed it "the Vomit Comet"). With a vacuum chamber aboard, a sail can be tested in an environment close to that of space, although the true test of a sail in orbit will, for NASA at least, take several more years.

Sail deployment depends largely upon the type of sail being used. While Salama has been working with spinning sails, a square or "three axis stabilized" sail requires a complicated external structure of booms to establish the proper tension. This makes the mass of the structure more and more of a burden as the size increases, so that square sails are optimal for sails in the 100-

meter diameter range and below. Salama believes NASA's early flight tests will be made with square sails because deployment is simpler. With a square sail, the sail membrane is guided into place by the boom structure, a complicated mechanical problem, to be sure, but easier to model than a spinning sail. For larger sails, however, spinning the sail seems the most efficient option. The spin persists throughout the cruise portion of the flight, with a fast spin at the outset to unfurl the sail materials. As the sail is deployed, its spin will slow down like the spin of an ice skater who extends her arms. Maintaining the final rate of spin keeps the sail under proper tension for the long cruise to its destination. "Circular spinning sails," says Salama, "are what we will use when we go to missions outside the solar system."

Getting even small sails deployed is tricky, as Russian scientists learned with Znamya. A second Russian attempt, Znamya 2.5, was made exactly six years after the first, on February 4, 1999, using a 25-meter sail-mirror with improved reflectivity and new systems for coordinating the beaming of sunlight to Earth. Unfortunately, the deployment failed when the sail-mirror became enmeshed with a communications antenna. Both the Znamya 2.5 sail-mirror and the Progress spacecraft to which it was attached burned up upon reentry into Earth's atmosphere. A third Russian sail-mirror has already been built, but there are no plans for launch.

Znamya was, it should be emphasized, a deployment experiment, not a free-flying sail. The mirror remained attached in both cases to the Progress supply ship from which it was deployed. Ground-based research into sail technologies continues at multiple NASA centers as well as in Europe, but neither NASA nor ESA has firm dates for flight-testing. At the same time, two plans to get a free-flying solar sail into Earth orbit are well advanced, one from The Planetary Society, the space advocacy group led by solar sail pioneer Louis Friedman, the other by Team Encounter, a privately funded venture based in Texas.

A 76 × 76-meter (249 × 249-foot) solar sail built by L'Garde will power what Team Encounter calls humanity's first starship, driving a 3-kilogram payload out of the solar system. Aboard will be messages, photographs, and even DNA samples, as many as 4.5 million, from fee-paying participants in the project. Launch by Ariane 5 rocket is planned for late 2005, with a precursor test mission the year before. After a 30-day orbital period for system tests, the sail vehicle will separate from the final launch stage, passing the orbit of Mars some 110 days later. Plans call for the spacecraft to achieve solar system escape velocity at a point between the orbits of Jupiter and Saturn, after almost two years of flight. The Team Encounter vehicle will cross the orbit of Pluto after thirteen years, traveling at a speed of 12.5 kilometers (7.8 miles) per second. At that rate, it would take 100,000 years for Team Encounter's sail to achieve the distance of Alpha Centauri.

NASA engineers working in late 2002 at Langley Research Center in Hampton, Virginia, successfully tested in a vacuum chamber the inflatable boom prototypes from L'Garde that will deploy the sail. These 54-meter (177-foot) booms support the sail in ways not dissimilar to the masts, booms, yards, and stays of an old square-rigger, but with a major difference: They have to be compressed to save space on launch. Once in space, the compressed booms are heated and inflated with nitrogen gas to expand them to their full length, extending them like a telescope. The booms, made of resin-impregnated Kevlar webbing (Kevlar is the same material used in bulletproof vests), are designed to become rigid after they deploy and encounter the cold of space, venting the nitrogen in the process. Deployments of small sail segments have been successful in tests conducted at L'Garde with no problems in ripping the material. Able Engineering has also done advanced work on sail deployment, addressing such issues as wrinkle management, for possible use in a NASA flight validation mission.

The U.S. National Oceanic and Atmospheric Administration (NOAA) has contracted with Team Encounter to study the feasibility of flying its precursor mission in a "polesitter" orbit, using solar pressure to hover the sail over the polar regions. A satellite in such a position could provide weather monitoring for areas not currently covered by weather satellites, and would also be useful for studying auroral phenomena and for providing communications to remote polar areas. Polesitters are another of those ideas that sprang fully formed from the head of science fiction writers. The ever-prescient Robert Forward took out a patent on what he called a "statite," or "stat"—any sort of non-orbiting spacecraft. By contrast, the geostationary satellites that NOAA now uses orbit at an altitude where the satellites can revolve around the Earth at the same rate that the Earth rotates. This altitude, approximately 22,300 miles, is a tenth of the way from the Earth to the Moon. It was science fiction writer Arthur C. Clarke who envisioned such satellites in a 1945 article in the magazine *Wireless World*. Clarke, then a radar officer in the Royal Air Force, failed to take out a patent on the idea, costing him who knows how many millions in the era of satellite television and communications platforms.

Forward's statite, described in a 1990 article for *Analog Science Fiction and Fact* and later in a story called "Race to the Pole," uses a solar sail placed over the polar regions "... with the sail tilted so the light pressure from the sunlight reflecting off the lightsail is exactly equal and opposite to the gravity pull of the Earth. With the gravity pull nullified, the spacecraft will just hover over the polar region, while the Earth spins around underneath it." Pole-sitters play a tricky balancing act between the Earth's gravity, the gravity of the sun, and the centrifugal force of Earth's orbit around the sun. Forward found that the minimum distance such a statite could remain from the Earth was 250 Earth radii (most communications satellites are 6 Earth radii out). That makes for a long round-trip lag time for communications, but it works for broadcast purposes, and NOAA believes it may work for meteorology as well.

The race to deploy a free-flying solar sail has also been joined by the Planetary Society, whose Cosmos 1 sail has research strands that extend back to the Halley mission. When the Halley sail was scrapped, JPL engineer Robert Staehle founded the World Space Foundation in 1979, developing a privately funded project to launch a solar sail demonstrator spacecraft. The design was sufficiently advanced to be the winning entrant for the Americas region in the 1992 Columbus 500 Space Sail Cup event, the aborted attempt to create an interplanetary regatta bound for Mars. Upon the demise of the World Space Foundation in 1998, its work was transferred to The Planetary Society, which has raised private financing through Cosmos Studios, a science-based entertainment venture created by Ann Druyan (the widow of Carl Sagan), as well as the A&E network and other sources, including the contributions of Planetary Society members.

Created without government funding, the Cosmos 1 sail is a real space mission. The spacecraft itself was built at the Babakin Space Center and the Space Research Institute in Russia, using a launch vehicle reengineered by the Makeev Rocket Design Bureau from a former intercontinental ballistic missile. The launch will take place from a submarine in the Barents Sea, off Russia's northwest coast.

Cosmos 1 takes advantage of existing Russian work on inflatable re-entry vehicles, a technology that will now be put to work on deploying the sail. Work began in the fall of 2000, with a suborbital test flight of the deployment system taking place less than a year later. The test failed, however, when the spacecraft failed to separate from the third stage of the Volna rocket. But planning goes forward on an orbital demonstration flight, with the 100-kilogram spacecraft due to be launched into a nearly circular orbit some 800 kilometers (497 miles) above the Earth. Onboard accelerometers will measure the effects of photon pressure, and a beamed microwave experiment using the Deep Space Network's Goldstone facilities will test the effects of this even more exotic means of propulsion.

Cosmos 1 is a 600-square-meter (6,460-square-foot) sail in the form of eight triangular blades made of aluminized, reinforced Mylar a quarter of the thickness of a trash bag (5 microns, or .000197 inches). In Earth orbit, the sail is deployed by using inflatable tubes filled with a foam that will become rigid once the sail is unfurled, leaving its blades extended and able to turn. This will allow ground controllers to vary the pitch of the blades to control how the sail is propelled by sunlight, much in the way a sailboat tacks as its skipper manages the wind. The blades can be rotated edge-on to the sun when no thrust is needed. The goal is to see whether photon pressure can be used to raise the orbit of the spacecraft, a proof of concept demonstration, but an experiment to measure ion flow around the vehicle will provide a scientific payoff as well. And, as we have seen, the spacecraft will test another propulsion idea: beamed energy.

Can a beam of microwaves from the ground have an effect upon an orbiting sail? James Benford thinks it can, and believes he has proven the concept in the laboratory. Benford, president of Microwave Sciences, Inc., in Lafayette, California, worked with his brother, science fiction writer and physicist Gregory Benford, on the design of Cosmos 1. Benford's laboratory experiments at the Jet Propulsion Laboratory using microwave beams to push a lightsail demonstrated accelerations of several times Earth's gravity. A separate laser experiment at Wright-Patterson Air Force Base in Ohio, performed by Leik Myrabo of Rensselaer Polytechnic Institute in Troy, New York, showed that laser light could impart thrust to a sail. In a press release from the Jet Propulsion Laboratory, Myrabo called these experiments "… the first known measurements of laser photon thrust performance using lightweight sails that are candidates for spaceflight." So laboratory work now confirms the hypothesis that a solar sail should function, and that beaming energy at it in the form of microwaves or lasers should cause measurable acceleration.

Benford's interest in beamed propulsion is long-term. An occasional science fiction writer, he has largely given up fiction to focus on microwave work not only for the Jet Propulsion Laboratory and Marshall Space Flight Center, but also for companies like Boeing, Raytheon, and Mitre. "I still write science fiction," says the wry Benford, "it's just that I call my stories 'proposals.'" Both James and brother Gregory were fixtures in the science fiction fan scene in the 1950s and 1960s when they edited a fanzine called *VOID*. It surprises no one who remembers that era that the two brothers remain deeply rooted in schemes as science fictional as driving a starship on a beam of energy, or that both should be working with practical, well-established physics to turn concepts in the pages of science journals into workable designs for future missions.

By 1999, enough had changed in the world of sail research to enable initial studies on such a prototype. "I realized that advances in materials had made it possible to do experiments on beamed propulsion," James Benford said. "Using microwave beams to move diaphanous sails that would be space probes. This hadn't been possible before because of two facts. The acceleration you can maintain on such a sail depends upon how hot you can let it get. And there is a built-in acceleration here on earth called g. So to do experiments on the flight of sails in the laboratory, one would have to reach accelerations of 1 g or more. Therefore we had to reach temperatures of approximately 1,500 to 2,000 degrees Celsius in practical materials."

Pure aluminum has a melting temperature of 660 degrees Celsius, well below the needed limit, and many metals are in the same range. But one material that survives high temperatures is carbon. Conventional metals go from solid to liquid to gaseous states as they are heated. Carbon, on the other hand, has no liquid phase. As it is heated, carbon "sublimes," going from a solid to a gas with no intervening liquid state, and this sublimation occurs at temperatures above 3,000 degrees Celsius.

It was when Benford learned that lightweight carbon structures could be built that he proposed his sail experiments to NASA, later demonstrating in a JPL laboratory that a 10-kilowatt microwave beam could lift and fly a 10-square centimeter sail in a vacuum chamber. "We saw the sail fly rapidly upward and strike the top of the chamber," Benford writes in a paper detailing the experiment. Accelerations of several gravities were observed in these tests.

The material a solar sail is made out of is a crucial matter. Because the sail itself must be extremely large in order to collect sufficient light, the sail material must be be extremely lightweight. And because the sail moves in a hostile environment, the material must be tough enough to withstand everything from micrometeorite strikes to huge temperature swings and the charged particles encountered along the way. Aluminized 0.5-micron Mylar, with a 100-nanometer (one thousandth of a micron) thick aluminum layer deposited on one side to provide a mirror-like surface for maximum reflectivity, is not ideal for longer sail missions because, according to JPL, it is degraded by the sun's ultraviolet radiation. Another material, called Kapton, is a candidate for missions in the inner solar system because of its ability to cope with ultraviolet, but it is not available in layers thinner than eight microns. That's workable for nearby space but far too thick (and certainly too heavy) for interstellar use.

And here the story turns to manufacturing techniques. For the critical measure for a solar sail's performance is "areal density," defined as the mass divided by the area of the sail. Areal density is normally expressed in grams per square meter. Cosmos 1's sail has an areal density of seven grams per square meter, while eight-micron Kapton has an areal density of twelve grams per square meter. Various methods are under consideration for producing sail films of lower areal density, including removing the plastic substrate and letting the 100-nanometer aluminum layer serve as the sail.

Further advances in sail acceleration could be achieved by cutting the aluminum layer down to 5 nanometers. While such methods would reduce the reflectivity of the sail surface, they would still offer substantial benefits when used with beamed microwaves. Perforating an aluminum sail provides a surface with even higher accelerations when pushed by microwaves, and Gregory Matloff has recently analyzed aluminum-mesh sails driven by sunlight, with a travel time to Alpha Centauri, in the best case, of 1,331 years.

The material of most interest to scientists, and the one Benford used in his laboratory work, is a carbon fiber material commonly called carbon-carbon. The double nomenclature refers to the fact that in this material, the carbon molecules are annealed to fuse them together in a structure called a microtruss, in which a core of carbon fibers is fused to a more textured outer surface. Carbon-carbon was developed under a JPL contract by a San Diego company called Energy Science Research Laboratories. Handed a sample of the stuff in Les Johnson's office at Marshall Space Flight Center, I was astounded by its almost non-existent weight. It was like a cloud of smoke made tangible. Seeing my amazement, Johnson explained the difference between carbon-carbon and materials like Mylar. "First generation sails will essentially be current polymers, which are like the glossy foil wrap you see at Christmas time. That stuff is fairly flexible; you can roll it up. Carbon fiber, on the other hand, while it is very lightweight, is also very rigid. Once you put a silver reflective layer on it, you'll have a great sail. Because it will maintain its shape, it's big, it's lightweight, and it's got the right thermal characteristics. It will demand something on the backside to dissipate excess heat, and rip-stops so any tear will run into a membrane to stop it. These are tricky issues; they'll take time to solve."

Two candidates for metallizing the carbon-carbon microtruss are silver or molybdenum (Leik Myrabo used molybdenum in his experiments at Wright-Patterson Air Force Base). Both allow for high reflectivity from the front side of the sail. And as we'll shortly see, a variety of even more exotic materials are under consideration for interstellar missions. But adding material to reflect photons also adds weight, which is one argument in favor of microwave methods. With pure carbon-carbon, light (with wavelengths down to the 10-micron level), goes straight through the sail. Whereas microwaves, with wavelengths up to several inches, bounce off even this diaphanous material, giving it a push. "Think of it like a screen door with a big grid," Benford told me. "The flies can come through but you can't."

For its part, the aluminized Mylar intended for early sail missions reflects light well, but because it melts at low temperatures and because of its sensitivity to ultraviolet, its long-term survival in space is problematic. And there are also differences in deployment. While Mylar can be deployed using various methods of spinning the sail, it must rely on mechanical systems to ensure that it does not tear or become wrinkled. But carbon-carbon stores mechanical energy when rolled up. As a carpet can be unrolled with a single push, so carbon-carbon can be rolled for storage aboard a payload into orbit and then simply released, allowing for initial deployment, which could then be assisted by other mechanical means.

Carbon-carbon fiber is remarkably light—it would take one hundred of its fibers to make a strand the size of a human hair—but it is also porous, so much so that it can be almost two hundred times thicker than the thinnest sail made from other materials and still remain relatively light. At 5 grams per square meter, carbon-carbon fiber has an areal density less than a third that of the plastic wrap that seals a pack of cigarettes. While it is twenty-five times lighter than standard office copier paper, it is stiff as a sheet of cardboard. Crucially, the material can be heated as high as 2,500 degrees Celsius without damage. That means a microwave beam could be projected at high power levels without damaging the sail, allowing a spacecraft to reach cruising speed more quickly on an interstellar journey.

One NASA report circulated within the Jet Propulsion Laboratory proposes using carbon-carbon microtruss fabric for a sail that would approach within 0.2 AU of the sun (some 30 million kilometers, or 18,600,000 miles, well inside the orbit of Mercury). The report envisions deploying a 1-gram-per-square-meter sail by wrapping it around a cylinder, and deploying it using centrifugal force. For reflectivity, the sail would use silver, titanium, or beryllium films connected to the carbon-carbon fabric, although advanced versions might have metals deposited directly onto the fabric. Such methods would be useful for imparting maximum photon pressure or for reflecting laser or microwave beams.

JPL's Hoppy Price notes that advances in sail materials could push interstellar precursor missions into higher prominence. "The In-Space Propulsion Program, in its last request for proposals, was requesting proposals for an advanced sail technology that could get down to less than five grams per square meter in areal density," Price said. "And when you get down to five, four, three, two grams per square meter, you can do a mission that goes in close to the Sun. You open up your sail, you power out from the Sun and you can reach up to 100 to 200 kilometers per second."

These speeds are far in excess of Voyager 1's 16 kilometers (9.9 miles) per second. But they're pushing up hard against the limits of our culture's perception of time. A 200-kilometer-per-second Alpha Centauri probe wouldn't reach its destination for well over 6,000 years. Obviously, we need more of a push than the Sun can give us, a source of continuous acceleration rather than a push from a light source that dwindles into the distance as we travel. If solar photons lose their punch as we cross the orbit of Mars, then we'll have to find a way to assist them through some form of beamed energy. We now know the physics of beamed laser and microwave energy. As some scientists say, "the physics works—the rest is just engineering." But we'll see what extraordinary engineering it is in the next chapter.

CHAPTER 6

OF LIGHTSAILS, RAMJETS, AND FUSION RUNWAYS

Terminology is always treacherous—one field's jargon contradicts another's unstated assumptions—but we must now make a distinction between solar sails and lightsails. The solar sail is pushed by photons from the Sun, while the lightsail is usually much larger and driven by man-made light, microwaves, or perhaps some kind of particle beam. This distinction isn't exclusive, though, because even before we've launched the first free-flying solar sail, the mission planners for The Planetary Society's Cosmos 1 have plans to demonstrate both photon and microwave propulsion on the same sail. Nor is size an absolute criterion, for solar sails can be vast in their own right. The early solar sail designs NASA studied for the Halley's Comet rendezvous reached 640,000 square meters, fully a half-mile to the side.

Better, then, to add these qualifiers: A lightsail is built in space and made of materials different from the plastic films being considered for near-term solar sail experiments. Not needing to be stowed for launch and later unfolded for use, lightsails can be built without the tough backing necessary to keep them from tearing on deployment, and can therefore be far lighter. "If you imagine a network of carbon-fiber strings, a spinning spiderweb kilometers across with gaps the size of football fields between the strands, you will be well on your way to imagining the structure of a lightsail," writes Eric Drexler, who has patented techniques for creating sails made of ultra-thin films. Drexler is a key figure in the development of nanotechnology and author of the seminal *Engines of Creation.* "Imagine the gaps bridged by reflecting panels built of aluminum foil thinner than a soap bubble, tied close together to make a vast, rippled mosaic of mirror. Now picture a load of cargo hanging from the web like a dangling parachutist, while centrifugal force holds the spinning, web-slung mirror taut and flat in the void." Thus one vision of a sail that could take us to the stars, a lightsail pushed by beams of energy.

Although solar sails seem more at home in warmer, light-filled interplanetary environments, they can be adapted for interstellar purposes. In fact, physicists Gregory Matloff and Eugene Mallove explored interstellar missions for solar sails in a classic 1981 paper, later refining them in their book *The Starflight Handbook.* The two envisioned a sail that would approach the Sun for a gravity assist, with a close pass by the solar furnace using an asteroid as a heat shield. With the sail unfurled and hidden in the asteroid's shadow, the spacecraft and occulting asteroid would swing as close as 0.01 AU (930,000 miles) from the heart of the Sun, skimming a scant 430,000 miles over the solar surface. At the critical moment, the sail would emerge from behind the asteroid, unfurling as acceleration builds.

In later studies, Matloff and Mallove would demonstrate solar sail designs that reach Alpha Centauri in approximately one thousand years. Depending on the sail materials developed by future technology and the velocity created by the solar pass, an automated mission might be possible whose transit time would be no more than several hundred years. Nor is the sail through when propulsion ends. Having achieved maximum acceleration, the starship could wrap the sail around its payload as a barrier to cosmic rays, perhaps deploying it again as a braking system upon arrival in the new planetary system.

Matloff and Mallove's starship would experience huge accelerations, up to sixty times that of Earth's gravity (60 g) if fully unfurled at perihelion, the moment of closest approach to the Sun. That puts the burden on sail materials, one reason the two proposed using cables made of diamond to attach the payload to the sail. As for the sail itself, Eric Drexler has shown that a technique called vacuum deposition can be used to create sails far thinner than the Mylar films proposed for proof-of-concept missions like Cosmos 1. Using these methods, metallic foils only a few hundred atoms thick have been created. Matloff and Mallove propose that a boron sail, with melting point of 2,600 degrees Kelvin, might be an ideal candidate for a mission this close to the Sun.

An interstellar transit of several hundred years, or even a thousand, would be an immense leap beyond what current technologies can achieve: the 40,000 years it will take Voyager 1 to drift within 1.6 light years of a star known as AC+79 3888 in the constellation of Camelopardalis, or the 296,000 years it will take Voyager 2 to reach the neighborhood of Sirius. Yet the idea that we could reduce transit times even more continued to push interstellar research through the latter half of the twentieth century. The breakthrough insight, moving well beyond currently available engineering but staying within the known laws of physics, was provided by Robert Forward in 1962. Working with the principle that the most feasible starship is the one that leaves its fuel behind, Forward began to think about pushing spacecraft with lasers. If the Sun couldn't continue to accelerate a sail much past the orbit of Jupiter, and if even the strongest gravity assist couldn't give you more than a few hundred kilometers per second, then finding a way to stream power directly to the spacecraft, pushing its sail with a high-intensity beam of light, might be the solution.

Forward had been working part time at the Hughes Aircraft Company, attending first UCLA on a Hughes fellowship in pursuit of a masters degree in engineering, and then the University of Maryland to work on a PhD in physics. At UCLA, he had begun compiling a bibliography on interstellar travel and communication, corresponding with authors and offering his own tips about ways to improve their work. At Maryland in 1961, he had set up the Interstellar Research Foundation, circulating ideas among fifty fellow interstellar enthusiasts, and calling for proposals to achieve flight to another star that did not use rockets. While the response was tepid, Forward in the meantime had made his own intellectual breakthrough, one drawing on the work of Theodore Maiman at Hughes Research Laboratories, who had shown how to make a functional laser in 1960. In the unfinished autobiographical essay he wrote near the end of his life, Forward recalled reading about the laser in a Maryland newspaper and thinking he had just found a way to send people to the stars.

"I knew a lot about solar sails," Forward wrote, "and how, if you shine sunlight on them, the sunlight will push on the sail and make it go faster. Normal sunlight spreads out with distance, so after the solar sail has reached Jupiter, the sunlight is too weak to push well anymore. But if you can turn the sunlight into laser light, the laser beam will not spread. You can send on the laser light, and ride the laser beam all the way to the stars!"

His enthusiasm growing, Forward wrote the concept up as an internal memo sent to the laboratory director at Hughes in May 1961. His first public statement of the notion came in a 1962 article in *Missiles and Rockets* called "Pluto, Gateway to the Stars" that was reprinted in *Science Digest* and *Galaxy Science Fiction*. The latter reprint is worth noting, given Forward's growing interest in science fiction and the strong connection between it and his scientific work. In fact, the most significant scientific article he wrote on laser-driven lightsails came two years *after* his novel *Rocheworld*, in which he laid out the essential elements that would guide his later research. *Rocheworld* tells the story of man's first expedition to another star, in this case, Barnard's Star, which has been chosen because earlier unmanned probes indicated the presence of planets, including the bizarre Rocheworld, a twin planet system held in the most delicate of gravitational dances. The title is drawn from nineteenth-century French astronomer Edouard Roche, whose work with gravitational forces on moons and planets led to the "Roche limit," the closest distance a moon can approach its planet without being destroyed. Forward's novel is stuffed with concepts notable not only for their ingenuity but their strict adherence to the laws of physics.

The starship of *Rocheworld* is driven by a sail fully 300 kilometers (186 miles) in diameter, with a 3,500-ton payload supporting some twenty crewmembers. How the ship—and the laser array that propelled it from an orbital station near Mercury—came to be designed is itself an interesting tale. Following his 1962 article (and in the aftermath of a subsequent examination of the laser-driven sail concept by George Marx in *Nature* in 1966), Forward had been in correspondence with science fiction writers Larry Niven and Jerry Pournelle. He invited them to meet him at Hughes Research Laboratories for a chat about physics. By 1973, when he met with the two writers, Forward still had found no way to stop his laser-pushed lightsail. Any interstellar flight using it would of necessity be a flyby. Niven and Pournelle nonetheless took away from the meeting the germ of their novel *The Mote in God's Eye*. "I had warned them that the light from the Sun was not strong enough to stop the sail," Forward recalled, "but, being science fiction writers, when they wrote *Mote in God's Eye*, they ignored my advice and pretended it would work."

Working on *Rocheworld* in 1981, Forward suddenly saw a way out of the stopping dilemma. If the sail could be built in two pieces, the large, ring-shaped outer portion of it could be cut loose as the ship approached the target star. The central portion of the sail would continue to pull the cargo and crew. Laser light from distant Earth would then be focused on the larger segment, which would bounce the light back onto the smaller segment, gradually slowing the ship over approximately a year's time as it approached Barnard's Star. In this excerpt from the novel, we see the starship Prometheus turning the inner sail (and payload module) around so as to receive laser light from the outer ring segment, called the ring-sail:

> As the central sail was almost halfway around, the ring-sail readjusted again and started to bring the rotation of the central sail and Prometheus to a halt. The teamwork of the four computers was perfect. The rotation stopped at the same instant the central sail was exactly one hundred and eighty degrees around. The central sail now had its back to the light coming from the solar system while it faced the focused energy coming from the ring-sail. Since the ring-sail had ten times the surface area of the central sail, there was ten times as much light pressure coming from the ring-sail than from the solar system. The acceleration on the humans built up again, stronger than before, but now it was a deceleration that would ultimately bring them to a stop at Barnard.

Of course, stopping at Barnard's Star is only half the battle for the human crew, who must now go on to explore the strange planetary system and cope with the fact that theirs is a one-way journey. They will spend the rest of their lives in space or on one of the planets around Barnard's Star. Whether a crew could be found to undertake such a mission is problematic, although the suspicion from this end is that there would be no shortage of volunteers.

In any case, robotic probes have fewer qualms about their own mortality, and Forward's dual-sail concept seems to provide them with a way to reach and explore a distant solar system. It is interesting to speculate what fictional use Forward would have made of P. C. Norem's 1969 paper analyzing ways of stopping a laser-pushed lightsail. Norem had the notion of launching his spacecraft on a trajectory that took it beyond the target star. As it moved through interstellar space, the craft would extend long wires and induce an electrical charge on the spacecraft. The charged vehicle, interacting with the interstellar magnetic field, would experience forces that caused it to turn in a large circle. Such a trajectory could bring the spacecraft to the target star from behind, at which point it could be slowed by the laser beam from Earth; theoretically, this process could be reversed for a return trip to the solar system. Probably the reason Forward chose not to use Norem's concept is that it would triple the total mission time. He had also developed doubts as to whether the interstellar magnetic field was strong enough to provide the turning radius needed for the spacecraft.

Forward's *Rocheworld* contains a lengthy appendix examining the technologies of the Prometheus mission. He went on to write up his concept in what may rank as the most provocative single contribution to interstellar studies ever made. "Roundtrip Interstellar Travel Using Laser-Pushed Lightsails" ran in the technical publication *Journal of Spacecraft and Rockets* in 1984, immediately spawning a swarm of proposals and counter-proposals to tweak and modify the enormous scale of the project. For Forward was a man who thought big. Engineering questions could be left for later: "The purpose of this paper is to show that interstellar flight by laser-pushed lightsails is not forbidden by the laws of physics," he wrote. "Whether it can be engineered and is financially or politically feasible is left for future generations to determine."

Maybe "big" is not the word for Forward's thinking; "gigantic" comes closer. He had noticed that Marx's earlier paper used an x-ray laser for propulsion, a position Marx was forced to take since he refused to consider a laser with a lens bigger than one square kilometer. Forward threw out the restriction. He knew that the laser system must be huge to create a beam that would remain tightly focused over such distances. He proposed that the performance of an interstellar laser would be improved by inserting a lens between the laser and the departing starship. He saw, too, that this focusing device, rather than being a single, solid lens, would be more practical if constructed as a Fresnel lens, a lens made out of concentric rings of, in this case, one micron-thick plastic film alternating with empty rings. The final diameter of this 560,000-ton object would be 1,000 kilometers, or 621 miles, allowing the laser beam to remain focused on the departing sail for distances measured in light years.

Such a lens is mind boggling, but the math seemed to demand it. The question is how tightly the laser would be able to focus light over extreme distance. The benefit of a laser in the first place is that, unlike ordinary light, it emerges in a tight, concentrated beam. At 4.3 light years, Alpha Centauri offered the nearest target for an interstellar probe, and at that distance, Forward found that a 1,000-kilometer Fresnel lens —one-third the diameter of the Moon—could focus light to a spot 98 kilometers in diameter, roughly one-tenth the size of his lens. The beam is, in other words, highly collimated, meaning it retains its shape even over interstellar distances, and at 4.3 years, the beam is still converging. Indeed, such a laser beam would not reach the size of its transmitting aperture—1,000 kilometers—until a distance of 44 light years.

So imagine this: A system of lasers and a combiner to pull their beams together into a single, coherent beam is placed in orbit near the Sun to receive maximum solar radiation for generating power. The lasers fire their beam to the huge focusing lens in the outer solar system—roughly 15 AU from Earth, between the orbit of Saturn and Uranus—which focuses the beam on the spacecraft. Forward imagined high-powered lasers in orbit near Mercury and gravitationally tied to the planet so the pressure of sunlight would not push the system out of position. The solar orbits meant that the lasers and focusing lens would be in constant motion in relation to the target sail, but Forward thought this could be resolved by having the spacecraft use a pilot laser that would allow for constant corrections in the outgoing beam. The alternative was to have the laser focus directly on the target star while using a beam-riding system aboard the spacecraft so that the vehicle made its own adjustments as required.

Edward Belbruno invited Forward to present these ideas at his 1994 "Interstellar Robotic Probes: Are We Ready?" conference in New York. Today, Belbruno chuckles as he remembers Forward in action. "Bob was definitely not constrained by engineering," Belbruno said. "These are huge lasers, but he knew that the physics worked. The disadvantage of a laser is that you don't get the momentum transfer you might get with a particle beam. But you can focus it better, and that's the bottom line. You want to accelerate things to high speed and keep that acceleration going. Because they take off so fast, the object is out beyond Mars within a few minutes. So you want to keep a beam on the probe so you can get the relativistic speeds and that means the beam has to be tightly focused."

Forward imagined several missions using this technology, beginning with a 1,000-kilogram unmanned flyby probe of Alpha Centauri with a 3.6-kilometer sail. Moving at just over one-tenth of the speed of light, the probe would reach the Centauri system in roughly forty years. The laser would be used to accelerate the probe for three years to a distance of 0.17 light years, at which point it would be switched off and the probe left to coast to its target. A second mission would involve a larger unmanned probe with rendezvous capability. Its two-part, 100-kilometer sail would operate just as in the novel *Rocheworld*, with the outer ring being used to reflect laser light back toward the inner 30-kilometer part of the sail and its attached spacecraft as it approached Alpha Centauri. The sail would receive a constant laser boost throughout the mission, reaching one-tenth of light speed. Forward's calculations showed that the inner sail could be brought to a stop in the Centauri system after one year of deceleration, with a total trip time of forty-one years.

A final starship version is the most exotic, involving a human crew with return capability. Using a 1,000-kilometer lightsail, the spacecraft would cruise at fully half the speed of light, fast enough to offer a 13 percent reduction in travel time for the crew, thanks to the relativistic equations of Einstein. After a 1.6-year acceleration, the starship would spend 20 years at cruising speed, or 17.4 crew years. Nearing Epsilon Eridani, the outer part of the three-part sail would be separated for deceleration, with a 1.6-year period required to slow down. Forward envisioned the crew using the light from Epsilon Eridani itself to power the remaining inner sail (he called this the rendezvous sail) within the Eridani system for exploration.

After 5 years of exploration, a 100-kilometer return sail would be separated from the center of the rendezvous sail, and the spacecraft (attached to the return sail) rotated to face the remaining ring segment of the rendezvous sail. A laser beam from the solar system that had been turned on 10.8 years previously, and left on for the needed 1.6-year acceleration period, would now illuminate the rendezvous sail ring, its power reflected onto the return sail for propulsion. Twenty Earth years later, spacecraft and crew would again be approaching Earth, where they would be brought to a halt by a final 1.6-year burst of laser power after a mission of fifty-one years in which they aged forty-six.

One proponent of the beamed lightsail is Freeman Dyson, who has largely given up on the idea of nuclear pulse propulsion in favor of this more exotic—and potentially less environmentally challenging—method. While acknowledging the massive engineering issues raised by beamed propulsion (he doesn't, in fact, believe that a laser lightsail will be launched in this century), Dyson sees laser and microwave options as scientifically feasible. And compared to the alternatives, they (and a method of "pellet propulsion" we'll investigate shortly) may be the only serious options known to our technology. Dyson, a key player in the nuclear-driven Project Orion and an early interstellar theorist, doesn't believe the nuclear gambit works on anything but the interplanetary level.

"Nuclear energy is too small," Dyson said. "You're using less than one percent of the mass with any kind of nuclear reaction, whether its fission or fusion. So the velocities you get with nuclear energy are limited to much less than a tenth of light speed, which is great inside the solar system but not outside it.

You might make a fusion starship, but it would take thousands of years to get anywhere. And by that time, there would be newer technologies that would overtake you wherever you are going."

But can lasers actually propel a spacecraft? We have little experimental evidence at present, although a maverick scientist with visionary ideas has caught Dyson's eye. Leik Myrabo is an Associate Professor of engineering physics at the Rensselaer Polytechnic Institute, in Troy, New York. He has been actively experimenting with a novel vehicle called a "lightcraft" for several years now. Myrabo was the investigator who demonstrated that a laser beam could be used to move extremely lightweight materials in the laboratory. In that work, which was sponsored by the Jet Propulsion Laboratory and performed at Wright-Patterson Air Force Base in Ohio, a sail five centimeters in diameter, made from carbon-carbon microtruss fabric and coated with molybdenum for reflectivity, was pushed by laser light and the effect measured. JPL's Henry Harris compared Myrabo's work to Robert Goddard's 1926 experiments with the first liquid-fueled rocket, which reached an altitude of only forty feet but set the stage for Vostok and Apollo. The point is apt because, like Goddard, Myrabo has now begun to turn raw theory into practice using available materials, in this case a high-temperature ceramic that can withstand its proximity to a blast of air heated to five times the temperature of the sun.

Putting that ceramic to work is a lightcraft, a stunningly simple Myrabo idea to propel a vehicle into Earth orbit using the power of lasers. As with the beamed lightsail, the beauty of this form of propulsion is that the energy source, in the form of laser equipment, stays on the ground. The back side of Myrabo's craft is a highly polished mirror. A pulsed laser beam is focused on the mirror from below, heating the air so rapidly that a blast wave is created behind the vehicle that pushes it upward. The plasma created by the pulsed beam—hot, ionized gas consisting of free electrons and ions—accelerates the vehicle and pushes it up to the speed of sound, at which point the laser light begins to convert intake air into more plasma that provides additional thrust. Like Orion, the lightcraft is now driven by a series of pulsed explosions, injecting onboard hydrogen into the plasma flow as the craft continues to climb. At the edge of space, well above usable air and moving at Mach 25, the lightcraft would use the rest of its onboard supply of hydrogen, heated by the continuing laser beam from below, as fuel for the final push into orbit.

While Myrabo has received government funding for phases of his work, his lightcraft have also benefited from the involvement of the private sector. Like the Cosmos 1 solar sail, funded largely by donations and by Cosmos Studios, lightcraft have been supported with a grant from the Foundation for International Non-Governmental Development of Space (FINDS), a nonprofit organization devoted to providing low-cost access to orbit. The physicist has been examining lightcraft—defined as any vehicle designed for propulsion by beamed power—since the 1970s. While at Teledyne Ryan Aeronautical in San Diego, he read two seminal articles by Arthur Kantrowitz in the journal *Astronautics and Aeronautics* that speculated on the use of lasers to propel payloads into orbit. Working at first on his own and later with a contract from the Jet Propulsion Laboratory, he refined the concept and later drew up his ideas in a 350-page report that would become his 1985 book *The Future of Flight*, written with Dean Ing. Today, Myrabo defines an explicit goal: a thousandfold reduction in the cost of getting a payload to orbit. Given that boosting a satellite on the Space Shuttle costs roughly $10,000 per pound, such a reduction would make for launch prices comparable to commercial airline tickets.

On October 2, 2000, a 4.8-inch prototype lightcraft designed by Myrabo's company Lightcraft Technologies reached a height of 233 feet. The tiny, futuristic vessel looked something like a shiny child's top, the sort that operates by pulling a string and setting it spinning, with a large upper cone partially covering a smaller, inverted cone emerging beneath. The vessel was stabilized for flight by spinning it at a rate of 10,000 rpm, at which point the laser pulses were fired. Powering these tests was a 10-kilowatt pulsed carbon dioxide laser, which had to be used with careful coordination through NORAD and the White Sands Missile Range in New Mexico (where the tests took place) to avoid illuminating satellites in low Earth orbit.

Myrabo's experimental lightcraft are small, but he has no hesitation in saying the design can be scaled up to launch everything from micro-satellites to manned vehicles. The size of the laser is the key. A 100-kilowatt laser—ten times the power used in these tests—will push a small lightcraft to the edge of space, and the eventual goal is to launch commercial satellites using 1-megawatt lasers. Myrabo figures that a 100-megawatt laser could power a manned suborbital flight like Alan Shepherd's Mercury-Redstone ride in 1961. And lasers of that size are well within range of current science. The Airborne Laser system created by Boeing, Lockheed Martin, and Northrop Grumman for the U.S. Air Force involves a megawatt-class chemical laser, which would be mounted in a Boeing 747-400 aircraft and used to destroy incoming missiles during their boost phase. Scaling studies for lasers up to the billion-watt level have been performed by the Strategic Defense Initiative Organization, according to Myrabo, who worked on a variant of the lightcraft idea for the group.

In Myrabo's view, lightcraft are the only short-term answer to inexpensive space access. "I don't see any other propulsion technology that's in the running," the physicist told me. "Single stage to orbit is hard to do with chemical rockets, whereas lightcraft are reliable because the engine is ninety percent on the ground. The system can be designed to run efficiently for decades, with downtime for maintenance just a few times a year. And if you're thinking about interstellar missions, you're going to need big power stations in orbit to provide the power for those laser-propelled lightsails. Building that space-based infrastructure is going to demand a quick way to space."

But even as we work on reducing the cost to orbit, the far more theoretical work on interstellar lightsails continues. NASA's Geoffrey Landis has studied Robert Forward's sail concepts extensively, with particular regard to the maximum operating temperature of the sail. It had been Landis who took the science of lightsail materials to its greatest height by analyzing a wide range of possible substances, starting with metal sails. Forward had proposed an aluminum structure for his sails, but Landis analyzed niobium, beryllium, and transparent films of dielectric (non-conducting) materials like diamond, silicon carbide, and zirconia. He was looking for ways to bring the craft up to speed faster, using a much smaller probe. Beryllium seemed a good candidate, although Landis would later reject it when he discovered it would heat to higher levels in the beam than he had originally calculated.

Today, Landis talks about diaphanous sails made of diamond-like carbon (a material closely resembling diamond), manufactured in orbit with a plastic substrate. Because of its high reflectivity and resistance to heat, such a sail could be accelerated to cruise velocity while still close to the laser source, making smaller lens sizes possible, and reducing the size (and cost!) of the sail. And construction of a film of diamond-like carbon seems possible by using the underlying plastic to shape the sail.

Once formed, the sail would be heated to make the plastic substrate vaporize. "You would be left with a soap bubble-thin piece of diamond film, the ultimate solar sail," Landis told me in his office at the Glenn Research Center. "As I envision it, you would probably have rolls of plastic material like rolls of saran wrap. As you deployed them, you would put the diamond film on top, making kilometers and kilometers of this material. And then you would put those pieces together in orbit, sublimating away the substrate, adding spars and ribs to give it shape. The sail is transparent. If it were under beam, you would see a bright spot because it would be tuned to reflect that light. And if the beam were turned off, you would see through it, except it still reflects that one wavelength. So it would look like a soap bubble in that wavelength. A bubble is transparent but you can see the colors."

Robert Forward had proposed using diamond laser sails as early as 1986, even as the notion of creating diamond films was becoming a reality. Industrial diamond has been produced for decades, compressing graphite in a hydraulic press for use in industrial applications like grinding lenses and machining metals. Today, chemical vapor deposition techniques have been proven to grow diamond film on non-diamond substrates. Although this technique is in its infancy, it is clear that diamond's many uses in cutting, machining, and managing heat will be useful in the electronics industry and elsewhere. Argonne National Laboratory—the direct descendant of the laboratory at the University of Chicago where Enrico Fermi created the first controlled nuclear reaction—has developed methods for making pure diamond films that may be used in microscopic motors, replacing easily worn silicon parts. Some of Argonne's films form continuous layers as thin as 500 carbon atoms. Forward had spoken of metal films some 200 atoms thick. Extrapolating the mass production of diamond films, including the likely development of nanotechnological manufacturing methods, a future lightsail is thus well within the realm of the possible.

Even so, questions remain as to whether lasers or some other form of beamed propulsion would be the most effective way to the stars. As early as 1984, Forward had begun to move beyond his own laser-driven lightsail to study an even faster concept. Drawing on unpublished work by Freeman Dyson, Forward conceived of Starwisp, an unmanned interstellar probe that would be driven by microwaves. Determined to create the lightest probe conceivable, Forward saw Starwisp as all sail, a wire mesh one kilometer in diameter weighing a total of sixteen grams, with microchips at each intersection in the mesh. Lighter than a spiderweb and actually invisible to the eye, the Starwisp probe would be accelerated at 115 times earth gravity by a 10-billion-watt microwave beam, reaching one-fifth of the speed of light within days of launch. As it passed through the Alpha Centauri system some twenty-one years later, it would use microwave power to return images from the encounter.

When Geoffrey Landis studied Starwisp in a paper on solar and laser-pushed sails for NASA's Institute for Advanced Concepts, he determined that Forward had been too optimistic about the effect of an intense microwave beam on the materials from which the spacecraft was made. "The error is in how reflective those tiny wires would be," Landis said. "If you try to fly the mission with wires that are as lightweight as Forward claimed you could make them, they would absorb microwaves; they wouldn't reflect them. In other words, if you tried to fly Starwisp, it would vaporize. It wouldn't last a microsecond. It would absorb all the power and vanish."

Heat is not the only problem with Forward's Starwisp concept. Because microwaves have wavelengths much longer than visible light, the system that delivers the microwave beam to the spacecraft must be proportionally larger than a visible-light laser. It must be, in fact, ten thousand times larger, which is why Forward's original Starwisp article called for a lens 50,000 kilometers in diameter, much larger than the diameter of the Earth (another of those small problems in engineering). Recognizing Starwisp's failings, Forward later re-thought the design, switching his attention to a "gray sail" made of carbon that would absorb sunlight during a close pass by the solar disk, achieving thrust by re-radiating infrared light.

Interstellar theorists can't let microwaves go because they offer some key advantages. They can be generated more efficiently than laser beams, meaning lower power costs. It is easier to build a large microwave aperture than a large laser aperture. And Forward believed solar-power satellites could supply the needed microwave output. Such satellites have been under study as ways of generating electrical power in space and beaming it down to Earth in the form of microwaves, where it would be converted back to electrical power for ground use. Starwisp would be able to use the beam from a satellite that otherwise was engaged in practical work supplying Earth's energy needs, somewhat lowering its cost.

Moreover, the technology to demonstrate the use of microwaves for propulsion is available today. Indeed, as we saw with James Benford's work at Microwave Sciences in the last chapter, microwave-driven sails have been tested in the laboratory and found to function, although with interesting side effects that require further study. The fact that a microwave sail does not have to be a solid film, but can be made in the form of a mesh, like Starwisp, is greatly in its favor, for a mesh profoundly reduces the mass of the sail (this was Dyson's key insight). "Of all the propulsion concepts explored," Microwave Sciences wrote in a final report on Benford's laboratory studies of microwaves, "beamed energy clearly works, i.e., needs no new physics, and has the most potential for near-term development."

But what if you could do away with all the spars and struts and deployment problems posed by a sail tens or even hundreds of kilometers in diameter, and still use sail-like methods for exploring space? The so-called magsail—a magnetic field acting as a sail—may be the answer. The original magsail concept grew out of Boeing engineer Dana Andrews's work in the late 1980s, in which he studied an interstellar ramjet concept that would use hydrogen ions from space as fuel for a nuclear-electric engine. Andrews's calculations showed that the amount of drag created by the system exceeded its thrust; the ramjet was not feasible. But Andrews and fellow engineer Robert Zubrin went on to consider the implications of this drag for another kind of propulsion. Why not create a magnetic sail tens of kilometers in diameter that would ride the solar wind to travel interplanetary space? Push it with a particle beam and it might be workable for interstellar missions. Intriguingly, the magsail can be generated from within the spacecraft, eliminating spars or supporting materials.

One version of such a "virtual" sail is being developed in the laboratory of Robert Winglee, a geophysicist at the University of Washington. Known as Mini-Magnetospheric Plasma Propulsion, or M2P2, the concept was originally funded by the NASA Institute for Advanced Concepts and is now under active study, with a prototype being tested at the Marshall Space Flight Center. If the name M2P2 rings any bells, it may be because of its resemblance to Star Wars robot R2D2. Winglee, an Australian with a puckish sense of humor, showcases the project on a Web site with Star Wars visual effects, including a scrolling text that mimics the movies' opening moments. To the swelling strains of the Star Wars theme, the viewer sees this:

> The year is Two Thousand One. It is a period of great technological innovation and experimentation. A group of scientists led by Dr. R. M. Winglee, working from a hidden base on the third floor of Johnson Hall at the University of Washington, have managed to secure a grant from NASA to test a prototype plasma propulsion device code named M2P2. Aboard a small space probe powered by M2P2, Dr. Winglee races to the farthest edge of the galaxy. His mission is to overtake the Voyager 1 spacecraft, and prove to the world that a coffee can, wrapped in a solenoid coil and named after a famous robot, really can go where no one has gone before.

One place where no one has gone before is the heliopause, that place where the solar wind runs into interstellar space. Voyager 1, launched in 1977, will reach the region somewhere around 2019, depending on the accuracy of our estimate of the heliosphere's size. Winglee's M2P2, if it had been launched in 2003, would leave Voyager 1 in the dust, reaching the same distance by the year 2013. And unlike both Voyagers, which have coasted for long years after their gravity assist maneuvers around the gas giant planets, M2P2 would use its electromagnetic "sail" continuously, gaining momentum from the flux of protons and electrons streaming out from the solar corona. That boost would allow a long period of acceleration as M2P2 departed Earth, diminishing only when the solar wind's effects ebbed in the outer regions of the solar system.

Injecting plasma into a magnetic field generated by M2P2 creates the sail, actually a huge magnetic bubble. Winglee talks about using a solenoid coil aboard the vessel to create a magnetic cloud approximately 15 to 20 kilometers in radius. A solenoid coil is a coil of wire wound in cylindrical form—thus Winglee's "coffee can"—that creates a magnetic field when direct current passes through it. In the case of M2P2, the solenoid produces a magnetic field approximately one thousand times stronger than Earth's. A 3-kilogram (6.6 pound) bottle of helium supplies material for the plasma, created by ionizing the gas with high-power radio waves. The size of the resulting bubble can be regulated by varying the amount of plasma injected near the coil. The results speak for themselves: The specific impulse of such a system would be ten to twenty times better than the Space Shuttle's main engine. Winglee thinks velocities of 50 to 80 kilometers (31 to 50 miles) per second are possible. Even higher velocities may not be out of reach; the solar wind itself pushes outward at speeds up to 800 kilometers per second.

Operating a solar sail, especially of the size needed for missions to the outer solar system (much less the stars) is a sticking point even for the designers of today's relatively small-scale sails. Winglee points out that while photons can impart higher momentum to a sail than the solar wind that pushes the magsail, the magsail can be readily scaled up in size without creating additional difficulties. Magsails of hundreds of kilometers for manned missions are feasible. "Problems of size aren't relevant to us," Winglee told me. "A sail stretched that far is only as intense as about a thousandth of the Earth's magnetic field. I call it a big bunch of nothing. You can move it around. Nature knows how to do this well—some coronal mass ejections can be as big as the Sun. We're figuring out how to do this ourselves."

Winglee describes a round-trip mission to one of Jupiter's moons to gather samples as being an ideal scenario for an M2P2 spacecraft. The solar wind remains powerful all the way to Jupiter, and Jupiter has a magnetosphere of its own that the spacecraft can use to slow its speed upon approach. The kinds of velocities possible for current magsail designs, up to 288,000 kilometers (179,000 miles) per hour, would get you from Seattle to New York in about one minute, or to the outer solar system in ten years. The hypothetical Jupiter mission would be a quick one, as opposed to the five or six years needed by chemical rockets, Winglee says. "We can get to a park [orbit] in the Jovian system in 1.2 years."

Numbers like that surely make M2P2 a candidate for future interstellar precursor missions of the sort NASA has placed on its roadmap of future development. Certainly Robert Forward appreciated the magsail, saying "I just love the audacity of that concept." It's no surprise that the magsail would catch Geoffrey Landis's eye. With the death of Forward, Landis has been focused on beamed propulsion concepts longer than anyone. In recent work, he has examined beamed-energy propulsion other than lasers, including particle beams that could theoretically push a magnetic sail to higher velocities than the solar wind.

The problem with lasers is that they offer extremely low energy efficiency, thereby demanding huge power sources. Forward had imagined lasers at the 7.2-trillion-watt level (scaling up all the way to 26 trillion watts for the deceleration phase), and even the more modest proposals that followed assumed lasers in the range of 25 to 50 billion watts. The huge lenses Forward imagined in the outer solar system also work against his concept—it is hard to conceive of the engineering required to build a 560,000-ton Fresnel lens in deep space. In fact, that kind of engineering presupposes so many spaceflight advances that other forms of interstellar propulsion might well become available before we could finish building the lens.

Landis ran the numbers on using a particle beam to push a magnetic sail. The idea is to strip electrons away from the atoms they orbit and then accelerate the positively charged particles, pushing them to speeds close to that of light. Because a charged beam "blooms" as it travels—particles with the same charge repel each other—the beam would spread, and thus would have to be neutralized (by adding electrons) to keep it together as it traveled through space before being re-ionized at the target.

A magnetic field has no heat limit because it cannot melt. That allows for faster acceleration by a powerful beam, and shorter travel times. The particle beam is far more efficient than a laser beam, and does not require a huge laser lens, while the sail itself can be enormous with no cost in weight—Winglee's M2P2 inflates the field by injecting plasma and needs no supporting struts and spars. All this makes the beam much easier to aim and control. Landis adds an interesting proposal to the mix: Use the particle beam itself to create the plasma, making a self-replenishing system that requires no material on board the spacecraft, unlike M2P2, which carries helium. "Within the solar system, a few kilograms of propellant every week is no big deal," said Landis. "But at interstellar distances, that extra mass becomes a problem. I see M2P2 as the beginning. It's a wonderfully elegant concept. I want to see them make these things and fly them to Pluto, or to a comet. What M2P2 can do in the solar system, we should be able to modify for interstellar work."

A magnetic sail has still another advantage. Once the spacecraft reaches its destination, the magsail can be used to slow it down by taking advantage of the drag against the stellar wind from the star. The magsail decelerates more efficiently at high velocities, and a more conventional solar sail could also be employed for the last phases of deceleration. Andrews and Zubrin went on to develop still another magsail idea, a hybrid concept called MagOrion, a spacecraft that used nuclear bombs set off behind the vehicle to create the plasma that would drive a magnetic sail. Such cross-pollination of ideas argues for the vitality of the magnetic sail concept, which is closer to flight testing within the solar system than any proposed interstellar technology other than the solar sail.

Variations on particle beams for propulsion continue to excite the interest of interstellar theorists, and not just for magnetic sails. Gregory Matloff and Edward Belbruno describe an Alpha Centauri probe using particle-beam propulsion, and in 1996 Matloff would expand this into a 900-year Centauri mission with a thin film of a spacecraft he called Robosloth. The craft would achieve initial acceleration by solar sail techniques close to the Sun before being positioned into the path of a particle beam. But even larger particles might be put to work. Physicist Clifford Singer had suggested as early as 1979 that a stream of pellets fired by an accelerator could actually drive a space probe. In essence, the pellet stream replaces the stream of photons from a laser or the charged particles in a beam.

Properly implemented, pellets could circumvent a major constraint: the spread of a beam of electromagnetic energy as it gets farther and farther away from the power source. Launched from within the solar system by nuclear or perhaps solar power, a stream of pellets (each a few grams in size) would not spread as it traveled, and the course of the pellets could conceivably be adjusted along the route. The pellets, traveling from 10 to 25 percent of the speed of light, would be vaporized into plasma upon reaching the vehicle, then redirected back as a plasma exhaust. Another variant would have the pellets captured by the vehicle and used as fuel in a fusion engine. There remains the engineering catch: Although pellets do away with the need for huge optical systems, they require equally vast electromagnetic launchers some tens of thousands of kilometers long.

Since the early 1990's, Gerald D. Nordley, a retired Air Force astronautical engineer, has worked off and on with the pellet propulsion concept. Nordley, who spent over five years in advanced propulsion research at the Air Force's Rocket Propulsion Laboratory near Edwards Air Force Base in California, has explored the concept of homing pellets and how they might push a star probe. The incoming stream of pellets must hit the spacecraft at distances of up to several hundred astronomical units. The spacecraft would ionize and reflect them back with a magnetic field to generate thrust. Nordley speculated that tiny, self-steering pellets could be created through micro- and eventually nanotechnology. These pellets, which might be shaped like snowflakes, would include their own direction and velocity control systems, as well as sensors and thrusters to home in on a spacecraft beacon, keeping the beam tightly collimated. Nordley thinks his self-steering, bacteria-sized pellets might weigh only a few micrograms, if losses due to occasional interstellar dust strikes are acceptably low. Singer's proposed pellets were relatively heavy (thus the need for such long accelerators). Lighter pellets could use shorter accelerators because it would be easier to send them at the high velocities needed.

To make such pellets, as well as the solar arrays to power the accelerators, Nordley turned to automated, self-replicating factories using raw materials from the solar system such as lunar regolith or asteroids. Such machinery could even be transported to the target star system, allowing each star visited by humans to build up its own infrastructure for further interstellar travel.

Nordley found his interest in pellets energized by an article that Dana Andrews and Robert Zubrin did for *Analog Science Fiction/Science Fact* that restated the magnetic sail concepts they had first voiced in their original paper in the *Journal of the British Interplanetary Society* (ironically, the peer-reviewed *JBIS* paper contained printing errors for key equations, so the science fiction magazine proved the more reliable source). If a spacecraft could be propelled by the pressure of the solar wind pushing against a magnetic sail, Nordley reasoned, why not create a plasma "wind" of your own? He subsequently discovered Singer's work on pellet propulsion, but by then he had extended the concept to include the magnetic sail, and saw that to turn his incoming pellets into plasma, he would need either a laser array in the starship to vaporize them, or, more elegantly, a self-destruct mechanism in the pellets themselves, which would be triggered by laser or radio waves as they approached the starship. The resulting plasma cloud would be reflected by the magnetic sail.

"Particle [pellet] propulsion has some fundamental physical advantages," Nordley added. "When they are reflected from the sail, if the speed of the particle beam has been programmed correctly, then the mass that's reflected from the sail is left dead in space with respect to the particle beam originator. What that means is that essentially all the energy used to accelerate the particles has gone into the kinetic energy of the starship, or whatever you're pushing. This makes for an efficient design, more efficient than any lightsail."

With homing pellets, Nordley's magnetic sail would be smaller than an Andrews/Zubrin magsail—perhaps as little as a hundred meters or less in radius—because rather than reflecting the relatively diffuse solar wind, it would work with a much more concentrated stream of pellets. His idea of adjusting the beam with self-steering pellets also significantly extends Singer's original concept. Singer had imagined the starship leaving laser-driven "collimating stations," designed to focus and re-direct any errant pellets along the route of flight. That principle is sound even today: So-called "optical tweezers" have been in use for two decades to move matter with light, especially in experiments involving molecular motors and other manipulations of small particles, including bacteria. But Nordley realized that rapidly developing nanotechnologies would make it unnecessary to use a separate infrastructure for adjusting the particle beam. By the time we're building starships, we should be able to create particles that know how to steer themselves.

Although given to speculation (in addition to his technical papers, he is the author of some forty science fiction stories), Nordley is reluctant to guess when an interstellar probe might fly. The technology is less a problem than the politics, in his view, but if the decision were made to do it, he thinks a robotic Centauri probe could be launched within the next thirty years, with a manned mission to follow in the early part of the twenty-second century. His one Arthur C. Clarke-style prediction: The first manned starship will reach Alpha Centauri in roughly the amount of time it took Magellan's crew to circumnavigate the globe, some three years. Three years as experienced by the crew, that is. Traveling at 86 percent of the speed of light, those aboard his starship would experience a time compression factor of two, meaning that half as much time would expire for them as for the people left behind on Earth. Nordley estimates half a year to accelerate, half a year to slow down, and about two years of transit time. A long but practicable journey.

Nordley's "smart pellets" provide a self-guiding beam to the stars. But it was Jordin Kare who developed the most outrageous of beamed-pellet concepts, a cross between a stream of particles and a lightsail. Kare's SailBeam concept grew out of his thoughts on the problems of beamed laser energy, specifically that the force exerted by the beam falls off as the spacecraft's distance from the power source increases. This made for enormous sails and the optical systems of Robert Forward that dwarf anything ever built by humans. But if you could beam photons, and if, as Singer and Nordley had shown, you could beam a stream of pellets, could you not also beam a series of small sails, each of which would become fuel for an outgoing spacecraft?

An astrophysicist and space systems consultant in San Ramon, California, Kare had realized that smaller sails pushed by a multi-billion-watt orbiting laser made scientific as well as economic sense. If you cut a large sail into smaller pieces and accelerate the pieces one after the other, the same amount of mass can be brought up to speed using a much less demanding optical system because the small sails can be accelerated much faster closer to the power source. Metal sails can't do this—they would melt under the intense power of the nearby laser beam. But the kind of dielectric sails considered by Forward and Landis should work. "I could push these sails really hard," Kare told me. "And now, instead of thinking in terms of taking a sail and dividing it into ten pieces, I realized I could divide it into a million pieces. I started doing the calculations and realized that this made sense as a propulsion system."

Kare first presented his concept to the Space Technology & Applications International Forum in 2002 (STAIF's annual meetings in Albuquerque are the initial proving ground for many advanced propulsion ideas). SailBeam works like this: These tiny beamed sails—"microsails"—serve as the "particles" sent to the spacecraft. The receding vehicle would use an onboard laser to vaporize the stream of sails into plasma behind the craft, using a magnetic field or, conceivably, a pusher plate to absorb the energy. SailBeam technology has several virtues, not the least of which is that it does away with the deployment and maintenance problems of large sail structures—the spacecraft is not coupled to the sails but driven by them.

In a report to NASA's Institute for Advanced Concepts, Kare found diamond film to be the most promising sail material, and worked out the logistics of spinning the sails for stability and guiding them along their course through a system of beacons. Ten 50-meter sail-tracking telescopes would be placed along the acceleration route, spaced several thousand kilometers apart. The entire acceleration path would total no more than 30,000 kilometers (18,640 miles). Kare's 1,000-kilogram interstellar probe would be accelerated by the pressure of plasma from the incoming sails. The SailBeam work synthesized earlier concepts: Kare had worked with Dana Andrews on MagOrion, which would trigger atomic bombs behind a magnetic sail for propulsion, and had collaborated with Andrews on a paper discussing how to use a magnetic sail for slowing down a spacecraft as it approached its destination star. His work incorporated Landis's studies of sail materials and Andrews and Zubrin's thoughts on magnetic sails. He would now propose melting a fleet of incoming sails into hot plasma, bouncing free ions off the starship's magnetic sail for thrust, and deploying the same sail for braking as the craft approached its destination.

SailBeam gives new meaning to the term "acceleration." In some designs, Kare's sails get up close to light speed in one-tenth of a second, although the final proposal uses the figure of three seconds to achieve the velocity. "I have a slide showing what the limiting acceleration would be for an ideal microsail," Kare said. "It's in the range of thirty million gravities. In that case, you go from zero to light speed in 0.97 seconds. I love showing that slide."

Kare's crossover methods, mating the beamed lightsail concept to pellet propulsion by making the pellets sails, have also led him to suggest another kind of synthesis, one that would use pellets to trigger fusion reactions, employing a so-called fusion "runway." The concept is impact fusion, which ignites the fusion burn by slamming two pellets of fuel together, one on the rapidly moving spacecraft, the other strung out in the form of a long line of pellets—the fusion runway—along the route of flight. As the spacecraft hits each pellet, fusion occurs. The threshold velocity for making impact fusion work is on the order of 200 kilometers per second, a fact that makes building an impact fusion factory on Earth an improbable goal. But in the realm of the outer solar system, not to mention the interstellar void, 200 kilometers per second is workable, especially with a gravity assist from the Sun.

In Kare's concept, pellets of fusion fuel would be sent out ahead of the spacecraft, which would then intercept them as it accelerated to speeds that light the fusion reaction, one pellet at a time. Imagine a doughnut-shaped spacecraft, a giant Cheerio or perhaps a cylinder, that carries its own supply of fusion pellets. When the spacecraft drops a pellet into the hole in its center, the onboard pellet meets the runway pellet and, as Kare says, "Bang!" If enough fuel pellets were strung in the right configuration along the route of flight, the fusion would push the craft to a cruising velocity of perhaps a tenth of light speed. At the end of its acceleration run, the spacecraft would be hurled along by a string of explosions going off at roughly thirty per second, each providing 100 tons of force. Needless to say, creating the fusion pellets and placing them precisely assumes a robust space-faring technology.

Although Kare built the case he presented to a Jet Propulsion Laboratory workshop around a one-ton interstellar probe, the beauty of the concept is that it is scalable. "The fusion runway doesn't care if you're working with a ten or a hundred ton probe," said Kare. "You just need more pellets. You don't need to build larger lasers. So it probably scales up better than most other schemes."

Unlike inertial confinement fusion methods, impact fusion doesn't require a spherically symmetrical pellet, or the application of force from all directions. Squeeze the fuel sufficiently from just one direction and you can achieve fusion. Theoretically, you could do this with a flat sheet of hydrogen, compressing it like flattening a sandwich. Or you could take two pellets of hydrogen and drive them into one another. If you can reach 200 kilometers per second, you can light the starship's main engines. One-tenth of light speed, a speed that would get you to Alpha Centauri in forty-three years, is 30,000 kilometers per second, within reach of the fusion-powered craft.

Kare calls his fusion runway vehicle the Bussard Buzz Bomb, after Robert W. Bussard, the man who suggested that a spacecraft built as a ramjet could suck up interstellar hydrogen as fuel. "Buzz Bomb" refers to the sound the vehicle would make (if you could hear it in space, which you couldn't), a series of staccato explosions much like the German V-1 pulse-jet rockets that terrorized London in the Blitz. When I brought the Bussard Buzz Bomb up to physicist Geoffrey Landis at his office at NASA's Glenn Research Center in Cleveland, he nodded enthusiastically. "It's a fascinating concept. The trick is that you have to be going pretty fast to begin with to catalyze the fusion. My thought on that is to do a close pass by the Sun." Landis turned to his desk and ran some quick calculations. "If you skim the surface of the Sun, you're going to reach 600 kilometers per second. You could line up your pellets to start hitting them as you go by the solar surface. That doesn't sound particularly unreasonable." The length of the fusion runway, Landis said, depends on how fast the spacecraft would accelerate. At one g, suitable for humans, a tenth of a light year of pellets would be required. For an unmanned probe that could pull one hundred g, the runway could be considerably shorter.

The Bussard Buzz Bomb offers up more than a few technicalities that will need some elegant engineering to solve. First, a magnetic nozzle has to be created that can contain and channel the resulting fusion explosions. And, of course, the whole concept depends upon each runway pellet hitting an equivalent ship-borne pellet precisely, which Kare envisions could be handled through a structure on the front of the vehicle to channel the runway pellets to a precise collision. Laser pulses to guide the pellets to their collision inside the spacecraft are the most likely way to proceed.

And as far as getting the pellets into space in the first place, they could be launched by small spacecraft that would drop pellets along their route of flight. Kare, who at the time was working with a group at the Lawrence Livermore Laboratory on the Strategic Defense Initiative, already knew how to build such vehicles. The anti-missile defense idea behind Brilliant Pebbles was to create space-based interceptors—small, self-contained spacecraft that could knock down a missile. A series of similar vehicles, each of them dropping fusion pellets along perhaps a million kilometers of runway, would over the course of ten years create a runway approximately half a light-day in length. The accelerating spacecraft would consume its pellets in ten days.

Thirty years ago, it would have been Robert Bussard's fusion ramjet concept, rather than Kare's impact fusion runway, that took pride of place among fusion enthusiasts. But technical reassessments have since cast doubt on what seemed to be the most elegant interstellar concept of them all, a spacecraft that not only carried no fuel, but needed to have no fuel supplied to it from Earth. Bussard first wrote about the idea in 1960 in an article whose title posed the issue precisely: "Galactic Matter and Interstellar Spaceflight." The Los Alamos-based physicist conceived of a scoop created by a magnetic or electric field. A forward firing laser would be used to ionize interstellar hydrogen, leaving its nuclei with a positive charge. The spacecraft's ramscoop would then suck in the charged hydrogen particles, feeding them into a fusion reactor that would convert the hydrogen into helium, producing thrust by tapping the same kind of fusion reaction that keeps the Sun burning. Like a terrestrial ramjet, the Bussard model would work only when the spacecraft attained sufficient speed, which meant that it had to carry enough fuel to reach approximately 6 percent of the speed of light to light the engine, and later, to decelerate.

As for the magnetic scoop, it would be enormous. Assuming one hydrogen atom per cubic centimeter as the density of the interstellar medium, a 1-ton probe would need a scoop fully 6,000 kilometers in size. Yet even this density seems high. A more realistic assumption of 0.1 atoms per cubic centimeter or less yields a scoop at least 20,000 kilometers in diameter, perhaps rising as high as 60,000 kilometers depending on hydrogen density variations in space. The fact that hydrogen atoms are the primary matter source is also problematic, given our current inability to drive a fusion reactor with them.

But if you *could* light that fire, the remarkable thing about the Bussard ramjet is that it would grow more efficient the faster it flew, making it, in theory at least, the ideal starship. When Carl Sagan examined the concept three years after Bussard's paper, he pondered an intake field some 4,000 kilometers in diameter, about the distance from Atlanta to San Francisco. Bussard believed that such a craft could accelerate continuously at one g, and Sagan ran the math. Total time to Alpha Centauri: three years. Ship time, that is, time as experienced by the crew, which would be considerably shorter than experienced by the ship's builders left behind on Earth. The effects of relativistic time dilation continue to multiply as the ramjet accelerates ever closer to the speed of light, but never, according to Einstein, reaching it. Soon things get, well, bizarre. Sagan found that with a constant acceleration of one g, the crew would reach the Andromeda Galaxy in a ship's time of twenty-five years, although over two million years would have passed on Earth. Explorations throughout the universe would become possible, though with no return to a familiar Earth. A Bussard flight is a one-way journey to the future.

More than a few of the scientists I talked to said the Bussard ramjet had been what fired their initial enthusiasm for spaceflight. And specifically, it had been a novel by Poul Anderson called *Tau Zero* that made them ponder the seemingly impossible. My own reaction had been similar; I still recall my paperback copy of the novel, read on a sweltering Missouri night with the stars ablaze and the crickets setting up a din from the deep woods. In Anderson's story (which was surely based upon the author's reading of Sagan), the starship *Leonora Christine*, bound for Beta Virginis some 32 light years from Earth, pushes ever closer to light speed because of a malfunction that makes it impossible to decelerate. Nudging up against c, the *Leonora Christine* experiences time dilation that is truly over the top, a journey that takes her through entire galaxies in what the crew perceives as mere seconds. "If there was any single book that turned me on to actually engineering interstellar flight," Jordin Kare told me, "it was *Tau Zero*."

Because Anderson went to great lengths to keep the physics as accurate as possible (at least until the mind-boggling climax), *Tau Zero* holds every bit of the fascination today that it did when first published. The term "tau" comes from the physics of time dilation. The closer the spacecraft comes to the speed of light, the closer the value of tau in the relativistic equation moves to zero. What that means is that an interval of time as measured on board the ship is equal to the amount of time an outside observer would measure with his own clock, multiplied by tau. The lower the value of tau, then, the more slowly time seems to run on board the vessel as observed by the outsider. At a tau value of one-hundredth, for example, crew-members would be able to cross a light-century in what seemed to them to be a single year, while a hundred years would have passed back on Earth.

Here we have reached one nexus that defines the fascination of interstellar flight, the astounding effects of high velocities on time itself. The allure of the Bussard ramjet is that it is the only design that suggests it is possible to attain such velocities. Anderson's novel is the definitive glimpse of what a Bussard ramjet might look like under power, and his description is worth quoting:

> The ship was not small. Yet she was the barest glint of metal in that vast web of forces which surrounded her. She herself no longer generated them. She had initiated the process when she attained minimum ramjet speed; but it became too huge, too swift, until it could only be created and sustained by itself . . . Starlike burned the hydrogen fusion, aft of the Bussard module that focused the electromagnetism which contained it. A titanic gas-laser effect aimed photons themselves in a beam whose reaction pushed the ship forward—and which would have vaporized any solid body it struck. The process was not 100 per cent efficient. But most of the stray energy went to ionize the hydrogen which escaped nuclear combustion. These protons and electrons, together with the fusion products, were also hurled backward by the force fields, a gale of plasma adding its own increment of momentum . . . The process was not steady. Rather, it shared the instability of living metabolism and danced always on the same edge of disaster . . .

Disaster indeed, but wondrous. A runaway Bussard ramjet, pushing up the long parabola closer to the speed of light at a constant one gravity acceleration, rolling up centuries, then millennia on the Earth it left behind while its crew lived normal human lifetimes, could reach the edges of the visible universe. Anderson's novel trades off the awe of a galaxy-hopping journey that takes people to places we've always assumed would remain faint imagery in telescopes.

More practically, a ramjet under control could power a robotic probe to nearby stars with absurd ease. And it wasn't just Anderson's novel that excited public attention. Four years earlier, when Sagan teamed with Russian astrophysicist I. S. Shklovskii to produce *Intelligent Life in the Universe*, his speculations on ramjet-propelled spacecraft became available to a much broader audience. Variations on the ramjet concept began to appear, including the so-called ram-augmented interstellar rocket, or

RAIR, conceived by Alan Bond (of Project Daedalus fame), which gathers interstellar hydrogen as it travels but uses fusion produced by onboard fuel supplies to accelerate the hydrogen into an exhaust stream. Another take on this technology is Gregory Matloff and Eugene Mallove's laser-electric ramjet, a craft that uses beamed power from a laser station in solar orbit to power up a reactor aboard the starship, with fuel provided by the ramjet's electromagnetic scoop.

What a splendid vision ramjets inspire. But both magnetic and electric scoop concepts have run into problems, with the likelihood of generating the kind of thrust that Anderson envisioned diminishing. Gerald Nordley told me that despite their early popularity, ramjet concepts have all but disappeared from serious interstellar speculation. "It's not a practical concept," said Nordley. "Not practical because of the very low amount of thrust you could get because of the density of interstellar hydrogen. Not practical because we don't have a good model for understanding how you fuse interstellar hydrogen and get energy out of it. And not practical because, like any ramjet, you're trying to add energy to an already energetic mass stream." Nordley uses the example of a child playing with a merry-go-round. Once you get the merry-go-round spinning, making it spin faster is difficult, because your arm can't move fast enough to add the extra energy. In a similar way, the ramjet is trying to add energy to itself when it is already moving at huge velocity. And there is that final "showstopper," to use a term I heard so frequently from interstellar theorists. As Zubrin and Andrews had demonstrated, drag would make the ramjet all but unflyable.

But the implications of an interstellar ramjet are too fascinating to ignore, and we can expect further study to see if there is some way to save the concept. And one thing is sure: Fusion, which has yet to be demonstrated in a practical way on Earth, has a long way to go before we can imagine missions anywhere close to the speed of light. The most productive concept to come out of ramjet theorizing is surely the wedding of magnetic sail to beamed propulsion. A magnetic sail pushed by beamed energy may be a viable technique for achieving speeds that would deliver a probe to Alpha Centauri within the lifetime of a researcher. At destination, a magnetic sail is our best way to slow that probe down, with perhaps a separate solar sail deployment at the end that can brake the vessel into Centauri orbit. If you had to bet on the thing—if the human race decided a fast probe had to be launched and was willing to commit the resources to do so within the century—this is where the near-term technology exists to make it happen.

CHAPTER 7

BREAKING THROUGH AT NASA: SCIENCE ON THE EDGE

Marc Millis zips through the western suburbs of Cleveland in his green Chrysler convertible, the top down on a benign day in early spring. In the front seat, hair blowing in the wind, is physicist and science fiction writer Geoffrey Landis. I am in the back, getting the full brunt of the airflow, eyes watering as I lean forward to eavesdrop on the conversation up front. These two NASA futurists, working out of Cleveland's Glenn Research Center, have their eyes on the issues we'll need to resolve to achieve interstellar flight, but just now they're talking about frontiers and human courage, and how the two relate to each other. It is just two months after Space Shuttle Columbia's fiery end over Texas.

"If a test pilot crashes at Edwards Air Force Base," muses Landis, "they name a street after him, and the next day someone else flies another mission to see what went wrong. With space, things are different. Every mission has to be a success; we can tolerate no casualties. It may be a cultural thing. Maybe we've grown too afraid of risks."

And I'm thinking Landis has it exactly right. It is certainly not the people in the machines who fear the risks; it is the culture that sends them. No matter what the mission, we seem to have no shortage of volunteers.

Millis parks and leads the way into a Kashmiri restaurant, where we settle in for lunch. The food is exotic, the talk more so. While Landis's work with lightsails and beamed propulsion has been on the cutting edge—some see him as the next Robert Forward, a rigorous scientist with a flair for dramatic prose coupled with an equally strong streak of shyness—it is Millis whom reporters seem to call for the latest news about the far edge of science. As head of NASA's Breakthrough Propulsion Physics project, Millis spent six years mapping out questions and organizing research that might one day give us a sudden, unexpected way to shorten those decades-long journeys to the stars.

NASA's Glenn Research Center, where I have spent the morning, is located just west of Cleveland's Hopkins Airport, a busy industrial area dominated by the huge NASA hangar, with a cluster of buildings, mostly of 1940s era architecture, stretching out behind. Millis's office, outfitted for three on the ground floor of Building 86, includes his crowded nook festooned with photographs. One in particular snags the eye, a shot of a thoughtful Millis on the bridge of the starship Enterprise, the result of his frequent dealings with Star Trek writers. I had spent the morning listening to Millis and fellow scientist Edward Zampino discussing, in addition to the challenges of interstellar missions, the background and recent woes of the Breakthrough Propulsion Physics project. Funding as of mid-2004 remained in suspension, though Millis thinks the situation may not be permanent. As with any bureaucracy, responsibilities have shifted within NASA, and different doors have to be knocked on to restore the project to operational status. The trick is to find the right doors.

Right now, though, Millis is talking about robotic interstellar probes and the problem of timing. In many quarters, lightsails beamed by lasers or microwaves remain the propulsion method of choice. But a sail, even one moving at a tenth of light speed, still takes decades to reach its destination. And the problem of when to launch such a star probe becomes acute. Millis refers to the dilemma as "Zeno's Paradox in reverse." Zeno of Elea (circa 450 B.C.) wrote of a race between the Greek hero Achilles and a tortoise. The tortoise told Achilles that it would win if Achilles would give it a head start. An incredulous Achilles demanded to know why any creature as slow as a tortoise would think it could win. But he conceded the race before it was run when the tortoise explained that each time Achilles halved the distance between the two, the tortoise would have advanced a bit further still, and so on ad infinitum. For Achilles to win the race was thus a mathematical impossibility. Lewis Carroll and Douglas Hofstadter are only two of the writers who have had fun creating new dialogues between Achilles and the tortoise.

In reverse, Zeno's Paradox looks something like this: Our society decides to build a robotic interstellar probe and launch it toward Alpha Centauri. No matter how fast we can propel such a probe, the passage of time will mean that in a few years, a faster method will be discovered. At some point, a probe launched later than the first one would pass it. But that probe would in turn be passed by another. If the problem sounds familiar, it is something like the dilemma faced by computer buyers—if I wait just a few more months, the price will have dropped even further, and the latest generation of microprocessors will be even more powerful than what I can buy today. At what point do I buy?

"Incessant obsolescence might be a better term than Zeno's Paradox in reverse," an animated Millis told me. "A long time ago, when I was working with Robert Forward and Robert Frisbee on the Interstellar Propulsion Society, some of us thought we should make an educational CD that would illustrate this point. We imagined using a hypothetical faster-than-light craft, and as you go on your journey, you pass all the predecessor craft that went before you—the Pioneers, the Voyagers, then a nuclear rocket, then a lightsail, then an antimatter rocket."

You re-create, in other words, a tale A. E. van Vogt spun in a short story called "Far Centaurus" that ran in the January 1944 issue of *Astounding Science Fiction.* "Far Centaurus" tells the story of man's first interstellar mission arriving at the Alpha Centauri planetary system, only to find it populated by Earth people who developed faster-than-light technologies not long after the first ship set out. They have even named the four inhabited planets after the four astronauts on the first mission. But in typical Van Vogt fashion, the story doesn't stop there. Through a nifty use of time-warping physics, the crew winds up back on Earth not long after they set out, to listen to their own words fifty years later reporting on their journey.

Now that's the kind of tale that would get the attention of a Marc Millis or Geoffrey Landis. A morning with them leaves this layman trying to wrap his head around concepts that are all over the map. The Vindaloo at the Kashmiri restaurant is spicy and rich, and the conversation now drifts to model making, at which Millis is an expert. He's just constructed a slot-car track that he has written up in a modeler's magazine, and we talk about the details of miniaturization. Landis, in his own way, has been something of a model builder, although his models flew. Before making his commitment to science, he launched intricately crafted, radio-controlled model rockets in international competitions. Now the kind of miniatures he is most engaged in are rovers, as in Mars rovers, vehicles like *Spirit* and *Opportunity* that take photographs and gather rock samples as they move through a Martian landscape he vividly brought to life in his science fiction novel *Mars Crossing*.

Landis has written stories about interstellar flight as well, but he thinks science fiction hasn't helped the serious study of such missions. "Science fiction has made work on interstellar flight harder to sell because in the stories, it's always so easy," Landis said. "Somebody comes up with a breakthrough and you can make interstellar ships that are just like passenger liners. In a way, that spoiled people, because they don't understand how much work is going to be involved in traveling to the stars. It's going to be hard. And it's going to take a long time."

I am spending the day at Glenn talking about interstellar propulsion and the strange physics that might make such journeys possible in years rather than decades. Maybe even faster, depending on the nature of the breakthrough—which could include the issues of causality that faster-than-light travel would introduce. Most of these ideas are, to put it in the best possible light, long shots. But it takes just one observed anomaly that stretches an otherwise sacrosanct law of physics in just the right direction to make things interesting. And as I am learning, you can bring the same rigor to the exploration of the improbable that you bring to any scientific inquiry.

Marc Millis is proof of that, a bearded, enthusiastic man who overflows with ideas, so that his words come forth in a torrent, concept intersecting concept. In his office that morning, I was reminded of the words of Lawrence Durrell, who wrote about "... a land which ... had given itself up to dreaming, to fabulating, to tale-telling." But Millis is one of the most meticulous scientists I have met on my interstellar journeying, and the tales he tells are grounded in known physics. His "Warp Drive When?" Web site soberly considers everything from wormholes to space drives, but like a jazz musician, he sometimes enjoys breaking into a quick intellectual riff, as on what you would see if you watched a faster-than-light ship returning from a journey to the stars. Imagine yourself standing on a space dock. The ship has made its journey and is about to arrive home.

Now imagine the scene: The image of the departing starship is still visible as it accelerates toward its destination. But because the ship is traveling faster than light, it arrives back at the dock before the image of its journey does. The ship thus appears at the dock with no visual warning. Now there are two images, the docked, actual ship and the image of the departing ship. But light from the ship's return also begins to arrive, the most recent light first. Now you see the actual ship along with two other images, one of the departing ship, one of the ship as it arrives.

And here is the Alice in Wonderland aspect of the scenario: The image of the ship's arrival is played out in reverse, the image seeming to back away like a movie played the wrong way in a projector. At the turn-around point, the two images—departing and arriving ships—meet. The ship ceases to be visible as a celestial object, and the only remaining ship is the physical ship in the space dock.

"Of course, this is making a very naive and errant assumption—I'm assuming that as it is traveling, the ship is still sending out light at light speed even if its going faster than the speed of light, so this is not true in the strictest sense," Millis says. "But I use it to illustrate something else. When you're talking about violating causality by going faster than light, what is it that is violating causality? Is it the object itself, or the image of the object? Images are images. If your image comes back before you leave, can that somehow stop you from leaving?"

I am still pondering the implications of this scenario as we leave the restaurant to return to Glenn. It is becoming clear to me that the creation of visual cues that stretch the boundaries of existing physics is a kind of investigative tool (Einstein's famous thought experiments are an example, though you can trace such "gedankenexperimenten" through history, from Newton to Schrödinger and Stephen Hawking). Thought experiments not only engage the intellect but open the door to different ways of viewing existing laws. And the motivation for gaining such insights has much to do with time.

Can we reach the stars in months rather than decades or centuries? If so, we can open them up to manned exploration. But going faster than light requires a new physics. To understand just one of the reasons why this is so, consider what happens at the level of the very small. Particle accelerators at Fermilab (Fermi National Accelerator Laboratory) near Chicago and CERN (European Organization for Nuclear Research) near Geneva are in the business of elementary particles. The only way to observe what is inside a proton, for example, is to accelerate one to enormous speeds and smash it into other particles to see what comes out, which is how antimatter is created. The faster the proton moves, the more likely the collision will produce exotic results. But push a proton close to light speed and a remarkable thing happens: The proton gets more and more massive. So massive, in fact, that at 99.9997 percent of light speed, the proton is a whopping 430 times more massive than when it is at rest.

The consequences for interstellar flight are daunting. Accelerating even a tiny particle close to light speed takes huge amounts of energy. Enough energy, in fact, that Fermilab often chooses to run its experiments late at night to minimize the effects on the public power grid. A fusion spacecraft accelerating to just half the speed of light before stopping again would require 7,000 times the mass of its payload in fuel. The greater the mass, the more energy required to continue acceleration. Einstein's equations tell us that attaining the speed of light itself is impossible because the mass would theoretically become infinite. It is clear that for propulsion systems of any sort, the speed of light remains a barrier, and that if a breakthrough ever comes that takes us beyond the speed of light, it will not involve rocketry but a "warping" of spacetime that we can exploit for travel.

Wormholes, warp drives, quantum tunneling—they are all the stuff of science fiction of the more sensational sort, the kind of fast and loose space opera that assumes the basic laws of physics do not apply or that simply ignores them altogether in search of a good story. Nonetheless, they are also considered with a certain regularity inside NASA. The genesis of the Breakthrough Propulsion Physics project goes back to 1990, when Millis and a team of volunteers at Lewis Research Center (now named Glenn Research Center) began brainstorming the idea of creating a "field-drive" propulsion system, a drive that produces force from the interaction of matter and fields. Just as gravitational fields accelerate masses, a spacecraft could induce a field around itself that would accelerate the vehicle. The completely unofficial group, styling itself the Space-Coupling Propulsion and Power Working Group, began a loose collaboration with scientists at other NASA centers, government laboratories, universities, and industry. An Interstellar Propulsion Society emerged for a time, with some of the leading theorists wrangling over a wide range of issues.

In May of 1994 the Jet Propulsion Laboratory in Pasadena hosted an Advanced Quantum/Relativity Theory Workshop that mulled over the possibility of faster-than-light travel. Several approaches were suggested for investigation: (1) cosmic wormholes exist and can be used by space travelers; (2) there is a realm of physics for which the speed of light is actually the lowest possible speed, as opposed to the highest; and (3) interactions with another dimension might make it possible to manipulate the space and time we know to achieve faster-than-light speeds. Millis was there, and so were many of the major names in interstellar theorizing at the time: science fiction writer and physicist Gregory Benford; Robert Forward; the Jet Propulsion Laboratory's Robert Frisbee; Geoffrey Landis; Washington University's Matt Visser; and Frank J. Tipler, author of the controversial *The Physics of Immortality*.

Given the vast chasm between known physics and the possibility of faster-than-light technologies, how can such a conference proceed? For such advanced subjects, NASA sessions have for years operated under the so-called horizon mission methodology, a structured process designed to spur new thinking. Rather than simply forecasting scientific breakthroughs, horizon missions are precise: Give a group of scientists a problem that is presently impossible to solve. If faster-than-light travel is the goal, then imagine it has already been done. Then develop the breakthroughs that would make that future possible, no matter how wild they might sound at present.

Once the goals are in place and the breakthroughs defined, the gaps in our knowledge become obvious. Discussing technical approaches to bridge these gaps can point to alternative understandings of physics. The method draws on Einstein's creative processes, which he folded together under the term "invention." If there is a gap between existing data and the axioms that would explain them, scientists have to find a way to bridge the gap. They do this by making a mental leap. Want to discover something about the nature of light? Then imagine yourself, as Einstein did in a *gedanken* experiment, actually riding on a photon. By doing so, you bypass traditional logic, but the thought experiment may yield entirely new axioms that can be tested in the concrete world of perception and data.

Thus Marc Millis's starship in the space dock. It is hard for the layman to imagine a visual mode of thinking that yields axioms that can explain the behavior of space and time. But Einstein's visual methods were matched not only by those of Richard Feynman and Stephen Hawking, but also by mathematician Roger Penrose, the latter observing, "Almost all my mathematical thinking is done visually and in terms of non-verbal concepts, although the thoughts are quite often accompanied by inane and almost useless verbal commentary.... Often the reason is that there are simply not the words available to express the concepts that are required. In fact I often calculate using specially designed diagrams which constitute a shorthand for certain types of algebraic expression.... This is not to say that I do not sometimes think in words, it is just that I find words almost useless for mathematical thinking."

All these scientists have found ways to bridge the chasm between raw data and theories that would explain them. So that when JPL's Robert Frisbee, the host and organizer of the

Pasadena workshop, introduced faster-than-light horizon missions at the 1994 gathering, he was able to point to the imagery of science fiction (a kind of visual scientific dreaming) as a palpable way of seeing new paradigms for physics. And it was surely with a sense of both whimsy and possibility that Frisbee alluded to the three famous laws of science fiction writer Arthur C. Clarke: (1) When a distinguished but elderly scientist says that something is possible, he is almost certainly right. When he says it is impossible, he is very probably wrong; (2) the only way of finding the limits of the possible is by going beyond them into the impossible; and (3) any sufficiently advanced technology is indistinguishable from magic.

Clarke's third law has attained the status of a mantra in some circles, both science fictional and scientific. Robert Forward adopted it as the title of a 1995 collection of science essays and short stories. And in fact almost all breakthrough propulsion methods can be so classified, from warp drives to wormholes connecting different parts of space and, potentially, time. Visualizing the non-tangible produces weird results, like tachyons, which are particles that do not violate the Einsteinian restriction against faster-than-light travel because they never travel *below* the speed of light, or "quantized" spacetime, which might allow a starship to avoid the uncomfortable light speed restriction by jumping from below light speed to above it without ever being *at* light speed.

The strangeness of the quantum world, in which distant particles that have interacted in the past seem to maintain a faster-than-light connectedness, challenges visual thinking to the utmost because of its radical departure from the world of experience. Yet in strangeness may lurk possibility, leading Forward to say in a paper to the workshop, "To this date, FTL travel and time machines do not seem to be *forbidden* by the known laws of physics (this, however, does *not* mean they are allowed)."

By 1996, when Marshall Space Flight Center was creating its Advanced Space Transportation Program (ASTP), the need for more such visionary thinking was apparent, and research on the frontiers of physics was opening new possibilities in propulsion.

Refinements of existing methods were being explored through programs within the ASTP. What would complement the picture was a program specifically designed to look beyond known methods to genuinely new technologies. Marshall accordingly chose Lewis Research Center to become the site of its Breakthrough Propulsion Physics project. Marc Millis would become its director, and it was he who presided over its first conference, held at Lewis in 1997. Although the BPP would have a troubled life, its never-robust funding suspended in the summer of 2002, the project was nonetheless able to encourage work in theories that played around the edges of known physics. These are areas usually investigated only on a part-time basis by individual researchers, who might or might not be aware of each other's work without such a clearinghouse.

BPP's 1997 conference was designed to push exotic ideas out into the open and look for the kind of critiques that only skeptics can provide. Some of the papers presented to the eighty attending physicists and researchers offered jaw-dropping titles: "Hyper-Fast Interstellar Travel via Modification of Spacetime Geometry" and "Ultrarelativistic Rockets and the Ultimate Fate of the Universe." The ubiquitous Forward was there, with "Apparent Endless Extraction of Energy from the Vacuum by Cyclic Manipulation of Casimir Cavity Dimensions." And so was Lawrence Krauss, chairman of the department of physics at Case Western Reserve University, who discussed propellant-less propulsion, and went on to tell a reporter from the *Cleveland Plain Dealer* that many of the concepts discussed at the conference simply wouldn't pan out. "Some of the ideas that are going to be discussed already violate known laws of physics," Krauss said. "Some that don't, like warp drive, might be possible in principle, but have recently been shown to require 10 billion times the amount of energy in the entire universe."

The author of *The Physics of Star Trek,* Krauss had spent a lot of time looking at far-out physics. Wormholes connecting different places and times, even reaching into other universes, are old hat in science fiction. But some of the newer concepts at the Breakthrough Propulsion Physics workshop were only beginning to find their way into the toolkits of science fiction writers. Of these, perhaps the most exotic was the electromagnetic zero-point field, in which energy might be extracted from the sea of vacuum fluctuations that make up the very substance of the universe. These fluctuations cause virtual particles and fields to wink in and out of existence, producing enough energy in a tea cup, as Nobel Prize-winning physicist Richard Feynman once said, to boil away all the Earth's oceans. Or drive a probe to the stars. The trick will be learning how to extract the energy, for the vacuum is filled with huge positive energies that are zeroed out by equivalent amounts of negative energy. Marc Millis would note in his report, "In pioneering work it can be difficult to distinguish between the crazy ideas that will one day evolve into breakthroughs, and the more numerous, genuinely crazy ideas. Even though many ideas proposed for this subject are likely to be incorrect, they can still be useful by provoking other, more viable, ideas."

Funded by NASA to the tune of $50,000, the meeting was attended by scientists and researchers from a wide variety of institutions. Among its participants were twenty-six representatives of private companies, eighteen university researchers, twelve scientists from government laboratories, and sixteen representatives from various NASA centers. Millis wanted to use the conference to define the technical goals of the then-nascent Breakthrough Propulsion Physics project.

To do so, he focused on what he considered a single, currently impossible mission goal: practical interstellar travel. Conference participants were able to identify three breakthroughs that would be needed to make it possible: (1) propulsion methods that reduced or eliminated the need for a propellant. This would imply creating motion through a variety of techniques including manipulating inertia or gravity, or by mining the interactions between matter, fields, and spacetime; (2) finding ways to go faster, which implies finding ways to move a vehicle close to the speed of light and possibly beyond it through manipulating the motion of spacetime itself; and (3) discovering ways to create the energy needed to power these vehicles.

Alternative physics can be a catchy subject for the media. While the conference attracted relatively little publicity in Cleveland, the national media weighed in with interest. "Warp drive. Anti-matter. Wormholes. Time travel. Reducing gravity.

Getting power from thin air. They all seem the stuff of *Star Trek*, *Contact*, and other science fiction movies," wrote the *Orlando Sentinel* in a story that went on to discuss in sober fashion why breakthrough physics is needed, and what form such physics might take. The *San Francisco Examiner* called the conference "... a development straight out of *Star Trek* ..." but was careful to weigh both the benefits and the difficulty of the attempt. "The press has been surprisingly accurate in dealing with us," Millis says, but perhaps there is no reason for surprise. Millis mixes his own enthusiasm with a practiced caution, and reminds this visitor repeatedly that things worth studying don't necessarily pan out. That said, the fruit of such study could be nothing less than astonishing.

What if you had a space drive, for example, that could manipulate the fabric of spacetime itself? A drive that was described, not in an old pulp magazine with a garish cover, but in a respected scientific journal like *Classical and Quantum Gravity?* Written by Miguel Alcubierre, a theoretical physicist from Mexico who is now at the Max Planck Institute for Gravitational Physics in Potsdam, the 1994 paper laid out a method for achieving faster-than-light travel. The Alcubierre "warp drive" gets around Einstein's speed of light constraint by pointing out that we do not know how fast the spacetime continuum itself can move. A spacecraft that could induce the spacetime around it to exceed the speed of light would not itself break the light barrier. It would, in a sense, be a hitchhiker on the fabric of space, going along for a superluminal ride. Alcubierre put it this way: "The spaceship will then be able to travel much faster than the speed of light. However, as we have seen, it will always remain on a timelike trajectory, that is, inside its local light-cone: Light itself is also being pushed by the distortion of spacetime. A propulsion mechanism based on such a local distortion of spacetime just begs to be given the familiar name of the 'warp drive' of science fiction."

The most common comparison made to describe the Alcubierre drive is of a surfer riding a wave. It is the motion of the underlying substance that drives the surfer forward. Another way to view it is that the flat spacetime inside the Alcubierre "bubble" is moving by destroying the space in front of the bubble while creating more space behind it, "... as if a local Big Bang were occurring at the rear of the spaceship while a local Big Crunch was occurring in front of it," in the words of physicist John Cramer. To discuss it in such terms demands that we accept Einstein's view that the three dimensions of space and the one dimension of time are not a kind of static background upon which matter moves, but rather a dynamic and malleable substrate that can be shaped by the energy within it. It is Einstein's general theory of relativity that puts the Alcubierre drive in play, while his special theory of relativity, which holds that no object can travel faster than the speed of light, does not appear to be violated by such an approach.

If Alcubierre's drive sounds like something out of Star Trek, that may be because the scientist got his inspiration from the show. His hypothetical spacecraft creates a distortion in spacetime that expands behind the vehicle while contracting ahead. A spacetime that can expand and contract seems exotic, but theorists like Alan Guth have speculated that the early universe underwent an expansion far faster than the speed of light, leading to the possibility of engineering a similar effect by human efforts. One problem with the Alcubierre drive, quickly noted by Jose Natario at the Higher Institute of Technology in Lisbon, is that such a vessel would become uncontrollable because its crew would overtake any signals they sent forward to control the wall of the spacetime bubble in which they traveled. Physicists Michael Pfenning at the University of York and Larry Ford at Tufts University went on to demonstrate a far more serious issue: The Alcubierre drive seems to demand more energy than is available in the entire universe to make it work.

The energy involved is not of the common sort, either. An Alcubierre drive requires the use of exotic matter that possesses so-called "negative energy" to power the drive. If you consider what we see in the visible universe as made up of normal matter with positive energy, then normal matter and exotic matter would be driven apart rather than together when acted upon by gravity. Far more than a mathematical construct, negative energy has been demonstrated in the laboratory by Dutch physicist Hendrik Casimir, who described its operation on parallel conducting plates placed close to each other in a vacuum. The Casimir Effect that he predicted was successfully measured in 1958, an indication that negative energy exists, with still more precise tests successfully conducted by physicist Steven Lamoreaux in 1996. The question for starship designers is: How do you produce the exotic matter needed to manipulate negative energy?

"We live in a world where energy is positive," Geoffrey Landis told me. "We observe that you can have more energy or less energy, but you can never have a region of space where the energy is less than zero. But let's hypothesize that maybe it doesn't have to be that way, and that there could be some space with negative energy or negative mass, which in many ways is equivalent. It would have some very odd properties. Negative mass does exactly the opposite from positive mass, so if you push one direction on negative mass, it goes the opposite way. The acceleration is the opposite of the direction of force. But it's gravitationally backwards, too"

Imagine something made of negative mass orbiting the Sun. Because it is gravitationally opposite to normal mass, the object should be pushed away by the Sun's gravity rather than drawn toward it. But pushing negative mass away causes it to move toward the object doing the pushing, so gravity still works. Bizarre for sure, but negative mass is not the same thing as antimatter, because antimatter still possesses positive mass and positive energy. Moreover, we know how to make antimatter. Negative mass is hypothetical, a fancy that rises from a set of equations. "Maybe this stuff is impossible," Landis added. "But it's not logically inconsistent."

And as for the vast energy requirements cited by Pfenning and Ford, maybe they're not required after all. Chris Van den Broeck, of the Institute for Theoretical Physics at the Catholic University of Leuven, Belgium, sees the possibility of creating a bubble of spacetime using warped space, one that is large on the inside but tiny on the outside. Van den Broeck calculates that such a bubble, no more than a gram, could contain an entire starship. Van den Broeck is able to drastically reduce the total amount of negative energy required for such a drive. "The spacetime and the simple calculation I presented should be considered as a proof of principle concerning the total energy required to sustain a warp drive geometry," wrote Van den Broeck in his analysis. "This doesn't mean that the proposal is realistic."

Impossible but not inconsistent. Proven but not realistic. On such horns hang the dilemmas of breakthrough physics. Millis thinks a ship using the Alcubierre drive would be uncontrollable because of a "causal disconnect" with the leading edge of the craft. No one aboard the craft would be able to turn the device on or off. "I am fully willing to entertain the possibility that these things are impossible," Millis said. "But when you start asking questions, they bring up other provocative questions, and show just how little we do know about things. On the Alcubierre drive, I would consider there is more value in the details of the debates rather than in whether or not this particular scheme is going to work."

The Breakthrough Propulsion Physics project was and—assuming funding is restored—will be a NASA concern, though one that taps the expertise of outside experts and consultants as needed. But one other program with a NASA pedigree likewise pushes advanced technology, if not as far as BPP, and that is the agency's Institute for Advanced Concepts in Atlanta. Both NIAC and BPP are think tanks, calling for proposals, setting up conferences, and trying to light a fire under research. NIAC, however, ventures far beyond propulsion to address everything from spacesuit design to antimatter containment. The Institute was purposely designed to push thinking beyond the NASA envelope by soliciting calls for proposals from independent sources in aerospace engineering, astronautics, and physics, encouraging their work through agency funding. No one who works for NASA can fund a project through NIAC, giving a green light to researchers in universities and private industry.

Run by the Universities Space Research Association, a nonprofit consortium with membership from ninety colleges and universities, all of which have graduate programs in aerospace engineering or space sciences, NIAC supports its charter through an active Web site (www.niac.usra.edu). The results are impressive: well over a hundred studies on topics ranging from biologically-inspired robots for space repairs to defense systems against asteroid impacts. Some of the most intriguing interstellar ideas we've examined, such as Jordin Kare's microsails or Terry Kammash's magnetically insulated inertial confinement fusion design, have been developed as NIAC projects.

Researchers receive $75,000 to fund a Phase I study, and work that seems particularly promising can lead to a Phase II study with additional funding. The organization normally funds fifteen to eighteen Phase I studies and five or six Phase II studies every year. Phase II studies—Robert Winglee's M2P2 design, for example—may go on to be examined for further development within NASA.

"Genius is in the generalities, not in the details," says Robert Cassanova, a genial aerospace engineer who took the helm at NIAC after retiring from Georgia Tech Research Institute in Atlanta. "Look at Einstein. The generalities of his theories were where his genius was. The details developed out of much analysis by many other scientists. Einstein was known not to be a good mathematician. His genius was being able to visualize an explanation of something in nature, in recognizing some general theory that would explain something. We want people to think about the possibility of doing things in a different way."

While the Breakthrough Propulsion Physics project was designed to look at how to push a spacecraft to fantastic speeds, the NASA Institute for Advanced Concepts has a much broader charter. While funding studies in antimatter rocketry and solar sail design, it has also branched into aeronautics, biology, earth sciences, and robotics. Is it possible to build a "space elevator" that would lift payloads directly from the earth's surface along a huge cable ascending into earth orbit, as in Arthur C. Clarke's novel *The Fountains of Paradise*? Can the caves of the moon be used as habitats for future explorers? Can a spacesuit be made skintight and flexible while retaining its structural integrity? All these and many more seemingly science fictional questions have been studied by NIAC. Something of its spirit emerges in a quote from Cecil Beaton that Robert Cassanova likes to cite. Beaton was a photographer famed for his portraits of well-known people; he also worked as a costume and set designer in Hollywood. "Be daring, be different, be impractical," he once said, "be anything that will assert integrity of purpose and imaginative vision against the play-it-safers, the creatures of the commonplace, the slaves of the ordinary."

But if science fiction provides the visual metaphors for what breakthrough science might achieve, how do we propose to bring those metaphors into the range of experiment? Various methods have been suggested for at least some faster-than-light possibilities. Wormholes that lead from one place in the universe to another are conceivably detectable by studying gravitational lensing effects, whereby the image of distant objects is magnified by intense gravity. Experiments to measure the mass of neutrinos may help to determine whether the neutrino is a tachyon, a particle that moves faster than the speed of light. Studies of cosmic ray showers may demonstrate much the same thing. For that matter, searching for neutrinos from a black hole may be productive, since only faster-than-light particles would be able to escape from inside a black hole.

And what if you could prove the existence of a kind of particle that comes out of the vacuum itself? An astrophysicist named Bernard Haisch, director of the California Institute for Physics and Astrophysics in Palo Alto, believes that empty space consists of what scientists now call a "zero-point field" that contains energy. The laws of quantum electrodynamics make it a logical conclusion that particles are continuously coming in and out of existence at all times, though their energy is unavailable for practical use because of their evanescent existence. So convinced is Haisch of the existence of these particles and energy fluctuations that he thinks the term "quantum vacuum" may be a misnomer. Haisch believes what we call a vacuum is really a "plenum," a word meaning "a space completely filled with matter."

Such a zero-point field may account for inertia, the resistance of an object to changes in its speed or direction. Inertia has long been a scientific puzzle. The more you try to accelerate an object, the more force seems to be required. It is as if a hand is pushing back in the direction opposite the desired motion. Is inertia an inherent property of matter, or is it what happens when the object being pushed interacts with some kind of field? Haisch would say that objects resist acceleration because they are pushing against the zero-point field "acting like a kind of electromagnetic molasses that gets thicker the more you accelerate," as he wrote in a recent article co-authored by physicist Alfonso Rueda. If this is right, then matter and energy owe their behavior to their interactions with a field that we, in our ordinary lives, are not remotely aware of, though we are invariably subject to its effects.

Working with Rueda, a professor at California State University at Long Beach, and Harold Puthoff of the Institute for Advanced Studies in Austin, Texas, Haisch developed this concept in a 1994 paper in the prestigious *Physical Review*, the journal of the American Physical Society, and went on to present new work at the 1997 Breakthrough Propulsion Physics conference. Like the cosmic microwave background radiation left over from the Big Bang, the zero-point field (ZPF) can be considered what Haisch and company call a "sea of radiation that fills the entire universe." But whereas the cosmic background radiation is weak, the ZPF is highly energetic, though not obvious to us because it is such an intrinsic part of the universe that we are simply unaware of it. It is by definition a field of energy that is isotropic—that is, identical in all directions. Since it is the same everywhere, there is no way we can be aware of its existence through our sense organs.

Curiously, moving through the ZPF doesn't help either, because if you are moving at a constant velocity, the field is what physicists call "Lorentz invariant," meaning that it cannot be detected unless you are accelerating through it. But Haisch, Rueda, and Puthoff established in their 1994 paper that an object accelerating through the ZPF causes a force to be applied to the object that is proportional to the acceleration, but acts opposite to that acceleration.

It is this effect that Haisch believes is responsible for inertia itself. Put an electromagnetically charged particle into acceleration through the ZPF and it will experience resistance. The view stands classical physics on its head, forcing us to reinterpret the idea of mass. In the Newtonian model, mass possesses inertia as an innate property, a resistance to acceleration that is observable but impossible to explain. Using ZPF concepts, it is possible to say that this field acts upon the electric charge inside matter to create the effect of inertia. "In other words," wrote the three scientists in a later account of their work, "the magazine you now hold in your hands is massless; properly understood, it is physically nothing more than a collection of electric charges embedded in a universal energetic electromagnetic field and acted on by the field in such a way as to make you think the magazine has the property of mass. Its apparent weight and solidity arise from the interactions of charge and field."

If we accept the zero-point field and its interactions with matter, how do we go on from there to build a space drive that will get us to Alpha Centauri? Gravity may provide a solution. Einstein based his general theory of relativity on the idea that the inertial mass of an object (its resistance to acceleration) is the same as its gravitational mass (its mass in a gravitational field). So if the zero-point field is what makes inertia happen, then it may also be responsible for gravity, an idea that was first proposed by Russian physicist Andrei Sakharov in the late 1960s. We can see gravity, then, as the result of the zero-point field being disturbed by the presence of matter. And if we accept the idea that both inertia and gravity are the product of electromagnetic interactions, then manipulating either of them to create exotic modes of propulsion becomes a possibility.

Propulsion systems that can manipulate inertia have a long history in science fiction, dating back to E. E. Smith, whose *The Skylark of Space* first took readers out of the solar system. That novel, published in *Amazing Stories* in 1928, seems to assume a space drive that overcomes inertia, as for that matter does Edmund Hamilton's *Crashing Suns*, another interstellar adventure that ran in *Weird Tales* the same year. In both cases, though, and in so much subsequent science fiction, the lack of inertia in the sudden bursts of acceleration that take the heroes throughout the galaxy seems to be an oversight rather than an intentional effect. There is no Campbell-like probing of the implications of controlling inertia in the opening days of space opera, although to be sure, Smith went on to discuss an "inertialess drive" in his 1934 novel *Triplanetary*, using it to attain speeds faster than light.

In more recent times, Arthur C. Clarke zeroed in on inertia in his novel *3001: The Final Odyssey*, in which Jupiter astronaut Frank Poole finds himself restored to conscious life a thousand years after his original, disastrous encounter with the seemingly malevolent HAL computer. Still learning his way around a strange new environment, Poole finds himself on an elevator ascending a 36,000-kilometer (22,400-mile) tower that connects to a ring-like structure in geostationary Earth orbit. Although the elevator moves upward at a huge velocity, Poole realizes he feels no sense of motion. His guide Indra explains (with a sly Clarkeian reference to ZPF theory): "I don't understand how it's done, Frank, but it's called an inertial field. Or sometimes a SHARP one—the "S" stands for a famous Russian scientist, Sakharov—I don't know who the others were." Manipulating such a field produces a drive that acts simultaneously on every atom of the body, thus circumventing acceleration effects. This is, as Clarke points out, not as improbable as it sounds, for its effects describe exactly how a gravitational field is perceived.

The "others" in the SHARP drive's name were, as Clarke says in his notes to the book, Bernard Haisch, Alfonso Rueda, and Harold Puthoff. Indeed, Clarke believes that the trio's paper "Inertia as a Zero-Point Field Lorentz Force" may one day be regarded as a landmark study, and assumes as much in his novel. "An 'inertialess drive,' which would act exactly like a controllable gravity field, had never been discussed seriously outside the pages of science fiction until very recently. . . . If HR&P's theory can be proved," Clarke continues, "it opens up the prospect—however remote—of anti-gravity 'space drives,' and the even more fantastic possibility of controlling inertia."

Robert Forward joined the debate on the zero-point field at the 1997 Breakthrough Propulsion Physics conference. Forward had already discussed engineering methods for putting some of this energy to use in a 1984 paper discussing the Casimir Effect. He wrote that the attraction between two parallel uncharged conducting plates in a vacuum seems to be the result of vacuum fluctuations of the electromagnetic field. Imagine two mirrors approaching each other in a vacuum. As they move extremely close, longer electromagnetic waves will no longer fit between them, so the total energy between the plates will be less than the amount pushing them together from the rest of the vacuum. This attractive force increases with decreasing distance, and at atomic-scale distances has actually been measured at tons per square meter.

In a 1996 study for the air force, Forward provided a list of experiments that might be carried out to test the concepts involved in the zero-point field. One possibility is the construction of an aluminum foil vacuum fluctuation battery, in which the Casimir force works against the repulsive electric field between leaves of ultra-thin foil. The battery extracts energy out of the kinetic energy of the motion of the leaves as they condense into a solid block, recharging itself when electric power from a separate power source is applied to the plates, causing them to separate once again. The concept pushes the limits of present-day precision manufacturing, and comes nowhere near the amount of energy found in an ordinary chemical battery. But, as with so much of Forward's work, it is an example of a principle at work, one that holds out the promise of discovering other methods that can generate more power.

Another way of studying the Casimir force is to examine the mass of an object affected by it. Would that mass have an inertia different from that of normal matter? The fact that such experiments are hard to envision—how you insert a mass between parallel, uncharged plates that are an infinitesimal distance apart is just the first of many problems—means that we need to develop a technology simply to explore these phenomena. One project funded by the Breakthrough Propulsion Physics project is in the hands of Jordan Maclay, CEO of a Wisconsin firm called Quantum Fields, LLC. A former professor of electrical engineering at the University of Illinois in Chicago, Maclay focuses not on parallel plates but metal boxes—he calls them cavities—no more than one micron to a side. Build the boxes right, using tiny springs at the limit of current manufacturing capability, and you have a tangible tool to study the vacuum pressure mechanically. Such microscopic measurements should help Maclay quantify the real behavior of this quantum vacuum.

The zero-point field remains controversial, to say the least, and there are those, such as physicist James Woodward, who dispute its explanation of inertia. Working at California State University at Fullerton, Woodward sees inertia not as the interaction of an accelerated object with a local field (like the zero-point field), but as the result of its being acted upon by all other objects in the universe, including the most distant ones (this is a restatement of Mach's principle, so-named by Einstein after the nineteenth-century physicist Ernst Mach). How can distant objects have instantaneous effects? One answer takes things in a truly strange direction. Perhaps pushing on an object causes a gravitational disturbance that propagates into the future. The disturbance causes distant matter to wriggle, causing a disturbance traveling backward in time that converges on the original object to generate what we feel as inertia.

The idea sounds bizarre, but no more so than many of the ideas of quantum mechanics, and the notion of waves from the future has been the subject of serious work by the likes of Richard Feynman and John Wheeler. Mach's principle, Woodward said in a telephone interview, is only common sense; when you say something is accelerating, the question becomes what is it accelerating with respect to? The only meaningful answer, he believes, is the distant matter of the cosmos. The corollary is that in an empty universe, there would be no such thing as inertia.

Another implication of Mach's work is that an object undergoing acceleration may experience transient fluctuations in its mass. Woodward's theories, investigated by John Cramer in a Breakthrough Propulsion Physics study, have interesting consequences for propulsion: These tiny variations in mass could, in principle, be exploited to push a spacecraft without any need to expel a propellant (think of the "impulse engines" of Star Trek). Woodward's experiments and those of Cramer continue, and as Woodward reported to the 2004 Space Technology and Applications International Forum in Albuquerque, his latest experimental results fit the theory to within 5 percent of prediction. "If the Mach effect really is there in the magnitude that prediction says it should be," Woodward added, "then building thruster-type devices based on the same principle as these experiments should be straightforward." It may even be possible, without violating the theory, to create so-called exotic matter with negative mass, converting everyday material by gravitationally isolating it from distant matter.

The gap between the desirable (an impulse engine that could exploit transient mass fluctuations) and the currently possible (building an apparatus to measure the tiniest effects of such fluctuations) is enormous. The same could be said for other experiments funded by the Breakthrough Propulsion Physics project, including superluminal quantum tunneling, in which photons seem to move faster than the speed of light, and claimed anti-gravity effects over a rotating superconductor. The effect being sought is no more than a wisp, pushing measuring instruments to the limits of their capacity, and human ingenuity into uncharted areas of engineering just to build the apparatus. No one seriously speaks of overthrowing Einstein and, with one experimental swoop, producing a faster-than-light engine. But Einstein did not overthrow Newton, either. Instead, his theories explained tiny anomalies in Newtonian mechanics, such as why the planet Mercury had a discrepancy in its orbit, or why the pattern of background starlight is shifted by the presence of the sun. The tiny, unexplained effect is what wakes a physicist up at night, with its tantalizing hint of an inviting door a bit ajar.

Walking through that door could lead you only to another familiar room. Or it might lead to a different place or time. If we accept general relativity, and thus the concept that gravity is actually the result of space being distorted by the presence of mass, then the idea of a wormhole is bound to arise. After all, bending space is analogous to folding a sheet of paper. Two marks that appear widely spaced on the paper can be overlaid on top of one another if the paper itself can bend. Out at the end of the theoretical limb, wormholes could conceivably be constructed, in Robert Forward fashion, by creating vast engineering projects like a ring of ultra-dense matter the size of Earth's orbit around the sun. Such a wormhole would open into what Forward called a "hyperuniverse," where spacetime exhibits different properties than our normal spacetime. Two such rings, each opening into such a hyperuniverse, might conceivably be connected if each were spun to speeds close to that of light and charged with massive voltages. Keeping the wormholes open could be left to negative energy, which presumably could be manipulated on the same scale as the matter that went into creating the wormholes in the first place. Travel between the two, assuming they connected into the same hyperuniverse (by no means a sure thing!) might be significantly faster (and shorter) than traveling through normal space.

On the other hand, any civilization capable of engineering on that scale would probably have found a more effective, and certainly less expensive, way to make its interstellar jumps. And while wormholes are not outlawed by our current conception of astrophysics, neither have any been observed. If they exist in nature, they still don't help us to get a probe, manned or unmanned, to the stars. After all, the first task is to get to the wormhole, which might be hundreds or thousands of light years away. The second problem: Who knows where it goes? A probe sent through such a wormhole might or might not return data. Wormholes, in fact, are so far beyond the realm of laboratory experiment as to fall into the realm of fantasy. Is there any way a scientist might investigate one?

Conceivably, says Geoffrey Landis, although the difficulties are all but limitless. First, consider how rare a wormhole is likely to be. Wormholes could have been created by the unimaginable forces at work right after the Big Bang, but they would also have closed almost immediately unless there were something to keep them open. Theoretically, a so-called cosmic string could do the trick—a one-dimensional extended structure that is like a break in the structure of spacetime. Landis calls cosmic strings flaws in geometry, and for good reason. If you started in the center of such a string and tried to move around it, you would return to the place from which you started in less than 360 degrees. A negative mass cosmic string could also conceivably exist, and if one did, and if it had wrapped itself around the mouth of a primordial wormhole, then the wormhole could have become stabilized.

Those are a lot of "ifs," but then, it's a big universe. Landis knows that a wormhole passing in front of a star would cause distortions to the visual field that follow a certain signature. We know from experiment that mass can curve space, which is why so-called gravitational lenses can form, in which light from a distant object is bent by a closer one. Just as images of the background stars were shown to be bent by the sun during the eclipse of 1919 in a crucial test of Einstein's theories conducted by Sir Arthur Eddington, so the image of a galaxy can be bent by intervening objects to produce unusual optical effects. The Hubble spacecraft has produced memorable images of a galaxy in this way, a single galaxy whose image is duplicated in the photograph.

A cluster of elliptical and spiral galaxies between the telescope and the remote galaxy has caused the bend in spacetime responsible for the lensing effect.

Wormholes, if they exist, should possess the odd property of seeming to display negative mass, the result of the forces of gravity being transformed as an object goes through the wormhole from the other side. And while normal, positive mass, the kind we are familiar with, can focus light, negative mass would have the opposite effect. Any light source behind the wormhole mouth should be de-focused into a kind of halo, diffusing in all directions. The actual wormhole would doubtless not be visible. But an image of it, if it happened to pass in front of a more distant star or galaxy, would have a characteristic signature. When the wormhole is directly in front of the light source, the halo would be formed. When the wormhole moves slightly to the side, a spike of light would result. Thus an occultation of a star by a wormhole should produce, first, a spike of light, then a halo, then another spike. "We published our paper on this," said Landis, "because people are actively hunting for gravitational lenses with spectrophotometers that can track these effects. We wanted to say, keep your eyes open for this particular signature. You probably won't find it, but if you do, it would be our first evidence that wormholes actually exist."

One form of wormhole that can never be observed is the kind that may lurk inside a black hole. Certain kinds of black holes, those that are spinning, for example, or those that are electrically charged, may enclose a wormhole, or so the Einsteinian equations imply. Going through such a wormhole into another part of the universe has become a science fiction staple, but it would take an intrepid astronaut to make the plunge. Because no light can escape from a black hole, there is no way to verify the existence of the wormhole at its core. And when you go through the event horizon of a black hole, the mathematics of general relativity make it clear there is no going back. In a bizarre way, the time and space axis become reversed—your direction toward the center of the black hole becomes your future, as inescapable as the forward arrow of time itself.

Thus Landis's short story "Approaching Perimelasma," in which an astronaut plunging through such a wormhole toward what he hopes is re-emergence at the star Wolf-562 finds instead a bizarre celestial display: "Instead of stars, the sky is filled with lines, parallel lines of white light by the uncountable thousands. Dominating the sky, where the star Wolf-562 should have been, is a glowing red cylinder, perfectly straight, stretching to infinity in both directions." The astronaut realizes he is sensing time as a direction in space. The lines of light are stars extending from billions of years out of the past into the future, the red cylinder the star Wolf-562 itself, its planets appearing as a braid around the star. Getting his character out of this predicament requires the ingenuity of a physicist who isn't afraid to speculate about humans tweaking spacetime itself.

A humanly manipulated wormhole is indeed indistinguishable from magic, but more than a few people believe in magic, just as many think unidentified flying objects are spacecraft from other worlds. It was in response to a single question from a reader that Landis wrote a short story called "What We Really Do Here at NASA." The question: "What are you NASA people doing with that UFO you have in the underground laboratories out at Glenn?" The story plays with the supposed saucer and the other outlandish technologies that cult theorists believe NASA already controls or is in the process of manufacturing. A hint: Aliens like to play cards, but they can cheat by using mental powers humans don't have. So one NASA employee each night has to take the cards home to be shuffled.

Marc Millis groans when I ask him about the Landis story, which appeared in the author's collection *Impact Parameter.* "I haven't read that yet, and I should. But maybe I don't need to—we get enough mail that I probably already know everything in it." One among many examples, this one dealing with wormholes: "Not really knowing the origins of what I saw, (see attached) I can truly say that wormholes are a reality, although short lived. The image that I am sending you is of a Non-Linear Space-Time Displacement Event, a wormhole by any other name. It found its way into my apartment, several times, to remove from my existance [sic], some personal object, only to return the object to my existance [sic] at a later time and in a different location. Truth is stranger than fiction."

Out of approximately one thousand unsolicited messages a year, Millis reports close to a hundred fit into what he describes as the "lunatic fringe." The challenge of dealing with controversial science was ever thus. On the one hand, you treat your subject using rigorous scientific procedures, assuring your peers that your study is serious. You cope with those scientists who believe the very investigation of certain subjects is folly, while encouraging skeptics to provide information on what it would take to change their views. Where, in other words, does the science need to be before scientists acknowledge those anomalous behaviors that demand further study? And as you proceed, you must protect your flank from those who, lacking all credentials, claim they have solved these problems using methods that are anything but scientific.

This may be thankless work, but any breakthrough that could open up the stars profoundly justifies the effort. It is not puzzling that both skeptics and true believers can move beyond science in an impassioned embrace of their viewpoints. Human nature has not changed within recorded history, nor is it likely to change any time soon. What drives researchers to tread these controversial waters is the same thing that impels people to dream. Dreams come not by choice but by necessity, a kind of metaphysical compulsion as well as a physically explicable activity of the brain. Interstellar flight is one dream that may come true. No one can say which breakthrough idea, if any, may evolve into a true star drive (my money is on Woodward). But whether the means be an inertialess drive, an anti-gravity field, or a space-warping Alcubierre starship, one thing seems certain: As long as there are stars, there will be volunteers.

CHAPTER 8

INTERSTELLAR COMMUNICATIONS AND NAVIGATION

At the Jet Propulsion Laboratory's Space Flight Operations Center one day in 2003, the communications traffic arriving from NASA's Deep Space Network was intense but not atypical. Any nation with interplanetary spacecraft to be monitored rents time on the network, whose terrestrial tracking stations pluck the data out of the sky and distribute them as needed. Twelve thousand dollars a minute will buy you DSN access, a nontrivial but essential price to pay for keeping a spacecraft healthy. On this sunny Pasadena morning, the Center was also tracking a number of missions run by JPL itself. The Galileo spacecraft was speeding away from an encounter with Jupiter's moon Amalthea, a digital display counting time upward from the event on a screen overlooking the room. Another timer counted down the launch of SIRTF, the Space Infrared Telescope Facility, which would take place some seventy-five days later. A continuously updating screen tracked the DSN's schedule, a marker for each antenna intersecting the colored lines marking different spacecraft on the chart.

From the glassed-in viewing area overlooking the Center, I marveled at the nonchalance of its operators, tweaking a dial here, tapping a keyboard there, sipping a Coke as the space probes spoke. This is the place where data from the probes Spirit and Opportunity announced their spectacular, bouncing arrival on Mars in 2004. I remembered the Pathfinder mission in 1997 as well, the backslapping and hand pumping when the long effort paid off, and later the precision deployment of the Sojourner rover. Yet Pathfinder, I was now learning, was also a symptom of a problem: Its data returned at the rate of 8 kilobits per second. Compare that to the 56 kilobits per second modem that now seems archaic (compared to cable or DSL) for Internet communications. Spirit and Opportunity had an easier time of it, able to send data back to Earth at speeds up to 1 million bits per second by using either the Mars Odyssey or Mars Global Surveyor orbiters as relays. Even so, this is only about one-tenth the speed of a broadband Internet connection.

The quiet routine of the Space Flight Operations Center belies the reality of what is happening to the Deep Space Network. The system is reaching the point where it can barely handle the loads being placed upon it. JPL's James Lesh points to factors as simple as the frequency allocations for deep space missions. In the X-band, a portion of the radio spectrum set aside for space missions, spacecraft are given an allocation of 50 megahertz. But that doesn't mean that each mission owns a 50-megahertz window on the dial. "If we are sending a mission, and the Europeans are sending a mission of their own, and maybe we've got two or three missions out there already, they're all using X-band," Lesh had told me in an earlier telephone interview. "They have to coexist. And if they're all at the same target, each of them potentially interferes with the others." Thus 50 MHz may be the allocation, but only a tenth of that is likely to be usable. The congestion was readily apparent in late 2003 as four spacecraft from three space agencies—the European Space Agency's Mars Express probe and lander; Japan's Nozomi orbiter; and NASA's twin rovers—converged on Mars, putting even more of a load on a system that was already tracking, among many other missions, the Stardust Wild-2 comet rendezvous and the Cassini mission to Saturn.

The thoughtful Lesh is chief technologist for JPL's Advanced Multiple Mission Operations System, which includes the Deep Space Network. His responsibilities thus include not only the groundside operations of the communications infrastructure, but the antennae and systems that fly on the spacecraft as well. He knows all too well what can happen when the system overloads, as was demonstrated on August 28, 1993. That was the day the spacecraft Galileo, crossing the asteroid belt on the way to its rendezvous with Jupiter and subsequent orbital mission, encountered an asteroid called 243 Ida. This was the second encounter between an asteroid and a spacecraft, the first being Galileo's 1991 flyby of the asteroid 951 Gaspra. Ida proved to be intriguing: it's an oblong 52-kilometer (32-mile) heavily cratered object with, Galileo discovered, a round, 1.5-kilometer (.93-mile) "moon" (subsequently named Dactyl).

Earlier observations had suggested that satellites orbited some asteroids, but Galileo offered the first proof. Its data was of obvious scientific importance, but the Ida flyby coincided with a crisis. The Mars Observer probe had stopped communicating with Earth as it neared the Red Planet, and the sudden flood of rescue attempts occupied the Deep Space Network's antennae. That meant the critical navigation images that would support the Ida flyby were reduced in number, a daunting problem given that Ida's exact position was not known. Fortunately, the first images of Ida, retrieved several days later, showed that Galileo's onboard camera had indeed captured it, the spacecraft passing within 2,480 kilometers (1,540 miles) to produce the first high-resolution image of an asteroid's surface. The story's happy outcome, however, does not diminish the scale of the DSN's problem. Much can go wrong when competing spacecraft vie for the use of such a critical communications resource.

An Alpha Centauri probe will run into more than just a spectrum allocation problem. Because radio signals, like all electromagnetic radiation, spread and fall off in intensity with the square of their distance, communicating with spacecraft is always a dicey proposition. A spacecraft twice as far from Earth as its counterpart sends a signal that arrives with four times less strength; at ten times the distance it arrives with one hundred times less strength. Those numbers don't create many problems for near-Earth missions, even if radio signals, traveling at the speed of light, do create a slight delay in communications with a Moon expedition or (more significantly) a Mars rover. But the problems begin to multiply as we push into the outer solar system because the spread in radio signals becomes enormous.

The Mars Pathfinder mission, for example, returned a signal that spread to hundreds of times the diameter of the Earth. By the time the much more distant Voyager's 23-watt signal reaches us, it has attained a beam width roughly one thousand times the Earth's diameter. That makes for an extremely weak signal, amounting to twenty billion times less than the power needed to operate a digital wristwatch, according to JPL's Lesh. And because the data rate depends upon the power of the signal, acquiring mission information can take a long time. At roughly 30 AU from the Earth, Neptune, the scene of Voyager's last planetary encounter, is practically in Earth's backyard compared to Alpha Centauri. At 270,000 AU, a spacecraft in the Centauri system would send radio information with 1/81,000,000 the amount of energy that Voyager could deliver from Neptune. The same spacecraft that transmits a 100-watt radio signal from Mars to Earth would need a terawatt (one trillion watts) to push data at the same rate from Alpha Centauri. Lowering the power available lowers the data rate, making for some challenging tradeoffs.

But challenges are what interstellar flight is all about, and power is only one of them. Communications from a probe orbiting Alpha Centauri would be one-way traffic, due to the distances involved. In fact, Earth-based controllers will probably be in active two-way communication with a Centauri probe only until it leaves the solar system, at which point it will begin relying upon its own error correction and data processing systems exclusively. The probe must be self-contained and self-healing, able to acquire its communications target (Earth) and beam a tightly focused signal back. Radio signals, moving toward higher and higher frequencies in recent times, have been the method of choice for communicating with interplanetary spacecraft. But as we'll see, interstellar distances favor optical methods using laser beams. That makes precise navigation and control a necessity, for a Centauri probe would have to keep itself locked squarely on its target, the distant Earth.

As I walked back to my car, dodging JPL employees headed for lunch, I pondered how two early starship designs had tackled these problems. The British Interplanetary Society's Project Daedalus probe was to use its burnt-out second stage engine, forty meters (131 feet) in diameter, as a massive communications dish for radio transmissions (with the engine unavailable for communications during the 4-year boost period at the beginning of the flight, Daedalus's first data would return by laser). Daedalus was also designed with internal communications systems that connected the main vehicle to the robot "wardens" used to maintain it, as well as the eighteen autonomous probes that would be dispatched when the spacecraft reached the Barnard's Star system. During the long coast phase, instrument platforms would be deployed up to 10,000 kilometers away from the main vehicle, making scientific observations that would be communicated back to the starship and thence to Earth by radio.

At its encounter with the Barnard's Star system, these interstellar platforms would be retrieved, while the autonomous probes would study interesting planetary targets. Because of the huge amounts of data they are collecting, these probes would cache their information for eventual transmission to the rapidly receding starship, which, moving at 12 percent of light speed, would have blown through the Barnard's Star system in a matter of days.

A starship called Project Longshot, designed at the U.S. Naval Academy, would use laser communication on an Alpha Centauri probe, deploying six 250-kilowatt lasers. Three of these would be placed outside the expendable fuel tanks for communications during the acceleration phase, and three attached to the starship for communications as the Centauri system was penetrated. Laser systems, which seem to pack the most signal into the smallest space, are only now beginning to work their way into the thinking of spacecraft designers, but work currently underway points to them as a significant part of planning for deep-space missions in coming decades.

The Project Daedalus designers assumed that their starship would use an enormous array of one thousand 64-meter antennae to supply the Earth-based part of the radio link, a creation called Project Cyclops that was planned as the core of an attempt to communicate with extraterrestrial intelligence. But Project Cyclops was never built. Instead, space communications today take advantage of the Deep Space Network's three sites, carefully chosen to provide enough separation to allow constant contact with spacecraft as the Earth rotates: the Goldstone complex in the Mojave Desert at the U.S. Army's Fort Irwin Military Reservation northeast of Barstow, California; in Spain west of Madrid at Robledo de Chavela; and 25 miles southwest of Canberra, Australia, near the Tidbinbilla Nature Preserve. Shielded from extraneous interference by mountains at each location, the DSN's huge, steerable dishes have become symbolic of deep space exploration, snatching signals as faint as the fall of a snowflake out of the void. The Interplanetary Network Directorate manages the DSN for the Jet Propulsion Laboratory, which in turn manages the operation for NASA.

How the DSN works with the Voyager 1 and 2 spacecraft gives us a taste of why communications with an interstellar probe are so problematic. The largest antennae are enormous parabolic dishes 70 meters (230 feet) in diameter, three-quarters the length of a football field. The network also includes 34-meter (112-foot) antennae and 26-meter (85-foot) antennae, but with the two Voyager spacecraft well outside the orbit of Pluto, these smaller dishes no longer have the muscle to send the spacecraft commands. Getting accurate transmissions through—roughly equivalent to making an ocean-spanning jump shot—is crucial for the Voyagers, since these tell the spacecraft when to transmit scientific data back to Earth, and what operating modes they need to switch to as they push toward the heliopause.

Signals from the DSN are transmitted to the Voyagers' 3.7 meter (12-foot) antennae via a 20-kilowatt signal in the S-band (a band already becoming obsolete). That's a bit less than half the power of an average AM or FM radio station, and five times less than the most powerful such stations. In turn, acquiring Voyager's signals takes global coordination. When Voyager 2 sent its dramatic pictures from Neptune in August of 1989, the feat was managed by combining the resources of multiple antennae, the twenty-seven-dish Very Large Array facility in New Mexico, the Goldstone complex, and the combined resources of the Parkes Radio Telescope in Australia and the Canberra site. The Galileo

Jupiter orbiter likewise used linkages between Goldstone and Canberra to synthesize a single "virtual" antenna that allowed for better signal capture and higher data rates, linkages that included up to five antennae from three tracking facilities. The improvement was substantial, three times the data return that would have been achieved with a single 70-meter dish. By adding arrays of antennae to advances in data compression and encoding, vast amounts of data could be teased out of a spacecraft whose main antenna had failed.

The much older Pioneer spacecraft have provided even more of a challenge for the DSN. Pioneer 10, launched in 1972, reached the end of its science mission in March of 1997, having become the first spacecraft to cross the asteroid belt and the first to make close-up observations of Jupiter. But the DSN continued to track the spacecraft as part of its studies in communications for interstellar missions. Now headed toward the red giant Aldebaran in the constellation Taurus, which it will reach in two million years, Pioneer 10 was last heard from in January 2003. The problem is not in the network but the spacecraft, whose radioactive power source has evidently decayed to the point that it cannot communicate. Pioneer 11, which made the first flyby of Saturn in 1979, was last heard from in 1995. Because its antenna no longer points at the Earth, any signal it may still be sending is undetectable. The next encounter for Pioneer 11 is a star in the constellation Aquila, the Eagle, which it will reach ancient and blind, but more or less intact, in some four million years.

Even though radio communication has been remarkably effective in reaching distant spacecraft, interstellar probes will require changing the ground rules. Until now, space communications have involved the use of ever-higher radio frequencies. Today's standard is the X-band, operating between 8.40 and 8.45 GHz for deep space work and between 8.45 and 8.50 GHz for near-Earth operations. Because the amount of data that can be transmitted varies with the square of the frequency, the X-band can support thirty-five times the data rate of the older L-band, which transmitted between 1 and 2 GHz. Higher frequencies, in other words, get you more bang for the buck, allowing you to focus the information into a much tighter beam. The Ka-band, from 31.80 to 32.30 GHz, is where the DSN is headed in the near future, an improvement that will increase network capability by a factor of four or five. The upgrade is sorely needed because the DSN is hard pressed to keep up with existing traffic, much less the load likely to be placed on it by future operations.

According to researchers like Cornell University's Martin Harwit, who analyzed the DSN in a recent issue of the journal *Science*, a change of paradigm may be necessary. Along with his son, Alex Harwit of Transparent Networks in Milpitas, California, and Joss Bland-Hawthorn of the Anglo-Australian Observatory in Sydney, Harwit has been advocating laser dishes in the near-infrared to handle data-rich missions. Radio can get us part of the way there—pushing radio wavelengths up to 40–300 GHz would ramp up the data rate significantly more and tighten the beam—but the ultimate effect would be achieved by going optical. A beam transmitted from an antenna spreads at a rate known as the diffraction rate, which is determined by the wavelength of the signal divided by the diameter of the antenna. Going up to the higher frequencies (shorter wavelengths) of optical communications means the resulting signal is much narrower. Lasers have not yet been widely used in space, but Harwit and company point to one recent test of the concept: The European Space Agency's Artemis satellite was able to create a laser link with the French satellite SPOT 4 in 2001. Although the data rate was slow, the principle seems sound. A laser-based communications system would use telescopes instead of today's deep space dishes, situating them in mountainous terrain with little cloud cover—places like Chile or Hawaii—for optimum viewing.

Astronomers are already familiar with the wavelengths the Harwits and Bland-Hawthorn are talking about. They routinely use them to study the skies, a fact that gives us a leg up on the technology since ground-based optical telescopes that could record signals are already in place. A laser signal carries vastly greater amounts of data, making the spectrum-crowding problem less severe. In fact, because laser beams are so narrow, they rarely compete for the same space in the spectrum the way radio signals do. And because it would operate at high frequencies and is more focused than radio, a laser system places fewer demands upon the spacecraft's internal power sources. This is significant because a whopping 40 to 70 percent of spacecraft power must now be dedicated to communications during peak periods. Another trump card for lasers: The size of the apparatus needed to receive and send a signal depends on the wavelength being used. Optical telescopes used aboard spacecraft can thus be significantly smaller than today's meter-size radio dishes, a savings that can be converted into larger payloads.

How a laser system could help an interstellar probe communicate was discussed in a paper James Lesh and two colleagues wrote for the *Journal of the British Interplanetary Society*. Noting that laser power levels have been doubling every year—a trend that shows no sign of abating—and that beam tracking and pointing to great levels of precision has now been demonstrated, the authors propose a 20-watt laser system with a 3-meter telescope only slightly larger than the Hubble Space Telescope as the transmitting aperture. Once in Alpha Centauri space, the probe would lock onto our Sun and use it as a pointing reference, its signals received by a 10-meter telescope in Earth orbit to avoid any absorption of the signal by the atmosphere. The Earth receiver would use optical filters to remove most of the incidental light from the Alpha Centauri system. "This is a system that is feasible right now," Lesh told me. "If we had a propulsion system that could get us to Alpha Centauri in ten years, or twenty, I am saying that the communications system is not a problem."

The emergence of laser communications is a key part of the infrastructure that will support an interstellar probe. Lasers also allow for data return that could change the way the public views space exploration. Right now, missions like the Mars rovers have to operate with a mere trickle of radio information. If conditions are right, JPL scientists can gather data at a rate that seems science fictional compared to today's cable and DSL Internet connections. The high laser frequencies allow light signals to carry far more digital information, but because the laser beam is tightly focused, it requires less power. With data rates a hundred times or more faster than a radio system's, a laser could operate with half the power of radio and take up no more than a fifth as much volume aboard the spacecraft.

In addition to communications, memory and computing limitations also have an effect on how missions operate. A Mars mission like Mars Global Surveyor might carry both medium- and high-resolution cameras. But the higher resolution devices can only be used occasionally because their images quickly fill up the available memory aboard the spacecraft. Lesh estimates that Mars Global Surveyor has been able to use its high-resolution camera to map less than one percent of the planet. Improved imaging techniques such as synthetic aperture radar, terrain-mapping radar, and hyper-spectral imaging—where images are broken into numerous bands in the ultraviolet, visible, and infrared portions of the spectrum to study soil composition and geology—are even more demanding than photographs.

The result: Today's space missions produce limited visual information. Think of the murky images of Neil Armstrong setting foot on the Moon, and compare them to the photographs from Pathfinder in the Ares Vallis region of Mars. The Mars images, taken with a stereo camera, were unquestionably better, but they flowed in so slowly that NASA was forced to string photographs together to achieve the effect of a movie, showing the little Sojourner rover moving about the ancient valley. The Spirit and Opportunity rovers were an improvement, using the X-band to communicate directly to Earth, and also able to relay data faster through a link to the Mars Global Surveyor and Odyssey orbiters. Even so, sending video is a difficult process. With laser systems, true moving video—and high-resolution video at that, up to IMAX quality—would be feasible. Scientists would be able to work with real-time imagery and explore planetary environments in true 360-degree wraparound perspective.

Those are powerful incentives, provided laser communications can get past its major hurdle: the accuracy needed to point the narrow beam to a waiting detector on Earth. The Jet Propulsion Laboratory estimates that a laser with a 30-centimeter antenna on Mars could send an optical signal concentrated to just one percent of the Earth's surface. Missing the target would mean the entire signal is lost. JPL has been experimenting with a laser package called the Optical Communications Demonstrator (OCD), which is designed to establish a link with a receiver by acquiring a beacon signal and steering the beam toward it. Having sensed the beacon, the OCD will lock onto it and adjust itself as the signal moves, mimicking in a terrestrial environment a spacecraft in its trajectory. In a series of tests beginning in 1998, JPL researchers established an optical link between southern California's Strawberry Peak, Lake Arrowhead, and Table Mountain, a three-way course spanning 46 kilometers (29 miles). With the OCD on Strawberry Peak, a receiving station on Table Mountain broadcast a beacon that the OCD successfully acquired. An in-flight system would supplement the usual Sun sensor and star tracker used to find the Earth with a laser beacon sent from Earth.

NASA's first dedicated optical communications telescope has been built at JPL's Table Mountain facility in the San Bernadino Mountains, housing a one-meter laser telescope. It will be used as a testbed for developing systems analysis tools and refining the acquisition and tracking methods used by future space missions. Down the road (perhaps by the end of the decade) is a network of 10-meter optical ground stations to support deep space missions, each augmented with a one-meter telescope to provide a communications signal and reference beacon to help spacecraft point the laser signal at the telescope. An early checkout of the optical network could come with a Europa Orbiter mission being considered by NASA as part of its plans for exploring the outer planets.

Creating a laser infrastructure is something the Jet Propulsion Laboratory has been working on for years. A major proof of concept came in December of 1992, when JPL's Optical Communications Group demonstrated precision laser pointing and tracking in an experiment called GOPEX—the Galileo Optical Experiment. The spacecraft in question was the Galileo Jupiter orbiter, launched from the Space Shuttle Atlantis in 1989. Galileo used multiple gravity assists to achieve the velocity it needed to reach Jupiter, passing Venus and moving past the Earth twice in its trajectory. As the spacecraft receded from its second pass of Earth, its angle allowed its onboard camera to see half the Earth in darkness. The GOPEX team conducted its experiments at sites near JPL and in New Mexico between 3:00 A.M. and 6:00 A.M. to take advantage of the dark background. By simultaneously transmitting pulsed laser signals from the 24-inch Table Mountain Observatory telescope and the 60-inch telescope at the Starfire Optical Range near Albuquerque, the team was able to achieve reception, creating the record for farthest transmission and reception of a laser beam, a distance of six million kilometers (3,725,000 miles).

A second proof of concept came with GOLD—the Ground-to-Orbit Laser Communication Demonstration. Conducted in 1995 using JPL's Table Mountain facility, the GOLD experiment transmitted an optical signal to the Japanese Engineering Test Satellite ETS-VI. GOLD was the first two-way, ground-to-space optical communications experiment, studying the performance of the system and the effects of the Earth's atmosphere upon the signal.

And it's not just NASA that is exploring laser communications. The European Space Agency will use a 60-kilowatt laser to test a link with a spacecraft in lunar orbit some time in 2004. The SMART-1 probe, launched in September of 2003, carries an infrared imaging camera that will attempt to pick up a laser signal from a ground station in the Canary Islands, assessing factors like how much blurring is caused by the Earth's atmosphere. The vulnerability of optical communications to cloud cover is one reason why JPL is deploying three atmosphere visibility monitoring stations, two in California and one in Arizona, to measure the intensity of starlight on the ground. Such measurements will allow more precise estimates of how the atmosphere can affect an incoming or outgoing laser signal. Even so, estimates are that three receiving stations separated by 500 kilometers (310 miles) would allow at least one of the stations to be available for laser traffic 95 percent of the time. The first operational optical ground station could be available as early as 2008, with space-based receiving stations supporting missions to nearby interstellar space within twenty years, according to JPL's former director, Edward C. Stone.

So powerful does the laser model appear that some scientists have begun to ask whether optical technologies may be the way an extraterrestrial race might try to communicate with the Earth. Charles Townes, who invented the maser (a precursor to the laser that produced microwaves) while a professor of physics at Columbia University, and who published the first paper on the "optical maser," or laser, had already acquired an interest in extraterrestrial communications by the time that paper appeared in 1958. After reading a groundbreaking article by Philip Morrison and Giuseppe Cocconi on the search for extraterrestrial signals, Townes suggested in 1961 that advanced civilizations might be identified by studying the optical spectrum, and followed this up with an article with R. N. Schwartz in *Nature*. The idea of visual communications with other worlds had precedents: In the nineteenth century, German mathematician Karl Gauss (1777–1855) had considered ways to signal the inhabitants of Mars with lanterns and mirrors, while astronomer Joseph von Littrow (1781–1840), director of the Vienna Observatory, suggested in 1840 that a 24-mile wide ditch be dug in the Sahara Desert, filled with a mix of water and kerosene, and set on fire as a beacon. The French poet Charles Cros (1842–1888) went even further in 1869 by noting that a huge mirror could burn numbers onto the deserts of Mars, a surefire way for Earth to be noticed by its putative Martian neighbors.

Looking for brief but powerful light pulses from nearby stars seems to make sense, given the amount of information that a laser can carry. But precisely because they are so focused, lasers have a downside. Any advanced civilization trying to signal the Earth would have to be deliberately targeting us. Radio communications might be much weaker but would spread, so that a signal would not have to be so precisely beamed. Nonetheless, scientists at the University of California's Lick Observatory, the University of California at Santa Cruz, and the Search For Extraterrestrial Intelligence (SETI) Institute are working with a pulse detection system that could seek out high-power laser pulses. Using light detectors that search for arriving pulses in nanosecond (one-billionth of a second) periods of time, the scientists hope to identify extraterrestrial signals that can be separated from the background starlight. This is tricky work, because false alarms can be generated by anything from cosmic rays to radioactive decay inside the instruments themselves. But unlike conventional radio searches, optical SETI is not swamped by terrestrial interference from a multitude of transmitting stations. Another optical SETI experiment is underway at Harvard University and Oak Ridge Observatory, using signal processors designed by SETI specialist Paul Horowitz and graduate student Andrew Howard.

Whether or not optical methods identify an extraterrestrial signal, their viability at carrying information has seized the interest of those who design interstellar missions. Ralph McNutt's "realistic interstellar explorer," discussed earlier as a precursor mission that could eventually lead to a robotic probe to Epsilon Eridani, uses an optical communications system for one-way transmissions to Earth. McNutt calls for orbital receivers because of the increasingly weak signal a receiver would have to work with. A decommissioned and retrofitted Hubble Space Telescope would make an ideal communications receiver, assuming an appropriate servicing vehicle can be found.

The space-based infrastructure for interstellar probes will not just involve a single optical receiver, however. As laser systems begin to link the planets together, they will invariably send data across an increasingly complex network of orbiting spacecraft, planetary rovers, and manned missions. Our growing sophistication at operating distant spacecraft will lead to a solar system-wide information exchange that scientists will be able to tap much like today's Internet. Laser-receiving stations in orbit around the outer planets will communicate their incoming data to network nodes nearby, from which they would flow onto the interplanetary network. A communicating system of probes, each with its own telescope for laser linkages, could double as an optical interferometer, a "virtual" telescope of unprecedented scale that could study planets orbiting around other stars and help determine future targets.

The first steps toward such an interplanetary Internet are already being taken, as I learn later that day in a small, windowless conference room in Pasadena. There, in a building tucked deep into the Jet Propulsion Laboratory's hilly campus, a discussion of what a space-based network would look like is underway. Vinton Cerf, inventor of the transmission protocols that drive today's Internet, stands at the whiteboard musing over the diagrams created by that day's work. In the room as well are the key players of the InterPlanetary Internet (IPN) project: JPL's Adrian Hooke, a veteran of Apollo, along with JPL colleagues Scott

Burleigh and Leigh Torgerson; Robert Durst and Keith Scott from the MITRE Corporation; and Intel Corporation's Kevin Ford. All these names are familiar to me because they appeared on an Internet draft document submitted to the Internet Research Task Force that I have been annotating for the last three weeks, looking for clues to the merging of digital networking with deep space communications.

The topic this afternoon is security—how do you set up an interplanetary network in such a way that this tempting hacker target is invulnerable to attack? Cerf, natty in a dark blue suit and vest, is writing on the whiteboard. His attire is in sharp contrast to the others, who favor a range of JPL-casual khakis, blue jeans, and the occasional T-shirt. The talk is punctuated by the clicking of laptop keys as Cerf draws an arrow from what looks like a picture of a bus to a nearby box. The diagrams are purple with orange overlays and black annotations occasionally pointed to by green arrows, a riot of intellectual color. I am reminded of the purples and pinks of Robert Forward's mathematical notes that I had examined in Huntsville. There is talk of delay tolerances, of certificates that guarantee the authenticity of a signal, and what would happen to a message whose certificate had expired. The irony is that the team's security specialist, Howard Weiss of Sparta, Inc., is not present. "We only talk about security when Howie isn't here," Cerf jokes.

What this team is talking about has caught NASA's attention. Why not network all the robotic spacecraft the agency sends into space? Doing so would lift some of the burden off the Deep Space Network because instead of every mission having to communicate directly with the Earth, orbiters and rovers could talk to nearby spacecraft, which could consolidate their data and schedule later transmissions to Earth. That would relieve the Deep Space Network of much of the communications overhead for essentially routine tasks. Retrieving data could become something analogous to setting up file transfers on today's Internet. Manned missions could use the Interplanetary Internet to download e-mail and keep up with news from home, while Earth-bound scientists could tap into data from rovers on Mars, Titan, or Callisto. Such an effort could become international in scope. A glimpse of a network like this came in February of 2004, when

JPL controllers routed a signal to the Mars rover Spirit through the European Space Agency's Mars Express orbiter, receiving return signals along the same path. NASA had been communicating with its rovers through its own orbiters, but this was the first experiment that involved spacecraft from other nations.

As the Internet grew on Earth, it was a collection of connected networks, all feeding their data into a common set of protocols so that each would be accessible. The InterPlanetary Internet is a step beyond that: Think of it as a network of Internets, a way of setting up clusters of communicating spacecraft and ground equipment, each using relatively conventional Internet methods among themselves in a planetary environment, and then feeding their findings through the long-distance network "backbone." Just as the terrestrial Internet uses a high-speed backbone that feeds into local conduits for message delivery, so the IPN will funnel traffic between the planets—and perhaps one day the stars—for delivery to local points or back to Earth itself. All this is being pondered by the IPN Special Interest Group, among whose members I now sit taking frantic notes. The work has been funded by the Defense Advanced Research Projects Agency and is essentially a bridge between the space community and the more Earth-oriented Internet.

If the IPN seems like a logical extension of today's computing, bear in mind that the existing Internet protocols were fine-tuned for the terrestrial environment. For one thing, optical fiber isn't available between the planets, so all communications are wireless, either through radio or, eventually, laser light. The Internet functions by using a set of protocols called TCP/IP: Transmission Control Protocol/Internet Protocol. Data is broken into small data envelopes, labeled for routing through the network, and reassembled at the destination. TCP/IP relies on fast data turnaround to handle its work, with many of its key tools being what network scientists call "chatty": The computers involved are, behind the scenes, exchanging data over and over again to work their way through a transaction.

Sending a file through File Transfer Protocol (FTP), for example, takes eight round trips of connecting data before the requested file begins moving, and FTP servers will often time out (cancel a communications session) after five minutes of inactivity. Now put that in the context of a moving target like Mars. Mars can be as far as 400 million kilometers (248 million miles) from the Earth—when the planets are on opposite sides of the Sun—or as close as 56 million kilometers (35 million miles). That's distant enough that round-trip radio times vary from 6 to 44 minutes. Then factor in the interruption of transmission as planets pass behind the Sun, for example, and for these reasons alone the conventional Internet protocols cannot be used on the interplanetary level, at least not for communications from Earth.

The answer is to invent a new set of long-distance protocols set up something like electronic mail. Data will be sent in the form of "bundles" that contain self-sufficient information—no back and forth handshaking required. A request for a file could contain authentication information to validate the recipient of the data, the location of the file requested, and where the file should be delivered. So think of the IPN as its creators do, as using the model of the Pony Express. The rider picks up a mailbag containing many letters—bundles—and delivers them to stations down the line. At each station—spacecraft—the bundle is accepted and then moved along to its next destination. There is no such thing as instantaneous delivery, only the sure knowledge that the bundle, no matter where it is currently being stored, will eventually reach its destination. Local operations can be handled using conventional Internet protocols—for example, the data dealings between a planetary rover and an orbiting spacecraft. In those nearby circumstances, a shared infrastructure can be built up that communicates data easily among its members, while bundling it for transfer to the long-haul network. A network like this fits in neatly with NASA's now-emerging idea of a Mars network, one that would connect microsatellites in orbit around the planet with a larger relay satellite to support surface missions.

But isn't an interplanetary network going to pose problems to the already crowded Deep Space Network schedule? Not at all, JPL's Adrian Hooke assured me. "The problem with the DSN is that it is a point-to-point network. Literally. You have big chunks of iron on the ground and you point them at a spacecraft. And you're dedicated to that spacecraft until you break that connection and move to another spacecraft. This just eats up time. You have to tear down the connection, decalibrate, and then move the antenna, recalibrate and, finally, turn on the transmitter. Going from one spacecraft to another takes a long time. One of our reasons for wanting a Mars network is that you can dedicate an antenna to a trunk communication to Mars and then allow different spacecraft on the planet to log on. Networking gives you the sharing of a common expensive resource for long-haul links."

Hooke's involvement in deep space communications is longstanding. Even as the terrestrial Internet began its rapid evolution in the early 1970s, Hooke and his JPL colleagues had begun to think of space communications in terms of data packets. So-called "packet switching" moves data in discrete bundles around the Internet and allows them to be reassembled upon delivery at their destination. NASA and the European Space Agency formed an international standards body called the Consultative Committee for Space Data Systems in 1982 that attempted to standardize how packetized data would flow up and down a communications link. Betting on a winner in the ongoing protocol wars, Hooke and crew extended TCP/IP to work with spacecraft. Space communications were, in a fashion, a kind of parallel Internet. Hooke remembers a meeting with Vinton Cerf in 1998, where the idea of bringing the two standards together took on an instant synergy, Cerf being a space buff himself. Since then, the idea has been hammered upon and refined at gatherings of the InterPlanetary Internet Special Interest Group, and extended to the entire issue of "delay tolerant" networking, whether the delay is caused by planetary distance or lack of tools here on Earth.

"We have parts of the Earth that are partitioned from the Internet because there is no connectivity," Hooke reminds me. "They may have their own local internet structure but they're not connected to the global Internet. The Sami people of Lapland are an example. Their trunk communications mechanism is a snowmobile with a cell phone. So they actually achieve their physical point-to-point transport there not by radio but by gasoline." To achieve total connectivity, tomorrow's wireless network on Earth may look much like the IPN being hammered out in this room in Pasadena.

As a network of Internets, the IPN is not compatible with the Internet itself. Its methods have evolved to suit the unique environment in which its data travels. The 500-bit data headers that make up the address information for Internet packets, for example, simply take up too much space for the bandwidth-precious IPN, whose developers have created a format for much smaller headers. Even so, systems on Earth will be able to translate between the two networks, allowing scientists access to interplanetary data from the Internet.

Of course with broadened access comes increased risk. The security issues addressed this day at JPL are of grave concern because hijacking an interplanetary network would be the ultimate coup for the kind of people who write viruses and computer worms. The network would have to operate under the assumption of "mutual suspicion"—each node on the Net would need to verify the identity of any other node with which it intended to communicate. That odd vehicle drawn on the whiteboard in the JPL conference room might be a spacecraft billions of miles from Earth or a truck that delivers information to a technology-starved village in Africa. In both cases, the questions are the same: Who are the data from, and how can the sender be authenticated? How do we keep data private when they can be so easily intercepted? The IPN's issues, it seems, are issues that get at the core of communication itself.

A fully wired solar system will resolve some of the communications and navigation problems that have plagued earlier interplanetary spacecraft. Mars will be surrounded by a constellation of satellites that can help rovers and, eventually, humans on the ground find their position. These satellites will also act as relays to handle communications with Earth, while performing complex data analysis that would otherwise have to be managed by a single spacecraft. Ultimately, such a network could allow data rates that would provide high-resolution video up to IMAX quality, once laser methods are in place. The returns for science will be immense and, not insignificantly, space exploration would receive a public relations boost as people find they can log on to check the weather on Mars or follow a robotic rover on Titan with an image clarity that all but puts the viewer inside the picture.

Navigation satellites are already a familiar commodity, given our increasing use of the Global Positioning System, or GPS, that compares signals from three satellites and measures them against a reference time signal from a fourth. The GPS system, twenty-four satellites in all, can be modified to an accuracy of mere centimeters. The European Space Agency plans a similar system called Galileo. But an interstellar probe is in the position of today's pre-IPN spacecraft. Without the luxury of networking, it will have to manage all data storage and communications needs internally, and with frames of reference utterly different from all preceding missions. Earth satellites can use the horizon of Earth as their reference frame, while interplanetary missions can compute the spacecraft's motion in relation to the Sun. Unable to rely on Earth-based controllers for the bulk of its navigation, an Alpha Centauri-bound spacecraft will need to derive its position and control its propulsion internally, locating and correcting its course with respect to other markers in the sky, whether they be stars or distant galaxies. Once beyond the solar system, the probe will be unable to use the Earth for reference, and the Sun itself will gradually become just one among many of the brighter stars.

We're looking, in other words, at a problem unprecedented in space exploration. In today's interplanetary missions, the movement of the spacecraft as seen from Earth can be converted into information about its trajectory. Combining a knowledge of how far the spacecraft is from Earth, what part of its velocity is directly to or from Earth, and its position in Earth's sky, controllers can determine how the vehicle is moving in its orbit around the Sun. Optical navigation extends the data available by using images taken by the spacecraft itself to view a target against background stars. The four types of data, acquired over and over again during the course of a mission, paint a picture of the history of the spacecraft's movements over time. By measuring the Doppler shift of the vehicle's signal, its velocity can be measured, a common strategy with interplanetary craft, whose acceleration or braking translates as a measurable shift in the radio waves returning from the craft. Meanwhile, a so-called ranging pulse can be sent to the vehicle, and the time of its return used to calculate its distance.

Optical navigation, in which a spacecraft measures the position of nearby targets against background stars, is a technique going back to the Mariner Mars probes and subsequent Viking missions. NASA's Stardust mission, launched in 1999, used an onboard camera to navigate to within 300 kilometers (186 miles) of the nucleus of comet Wild-2 in early 2004, trapping particles from the comet that will eventually be returned to Earth for analysis. As part of the planning for its cometary encounter, Stardust took images of guide stars at the same point in its orbit that it met the comet, using multiple photographs to help it perform the delicate navigation needed on the approach.

But an interstellar probe will have to create a new coordinate system. The navigation problem seems on the surface to be relatively simple, but this is out of analogy to the night sky to which we've become accustomed. We know the planets are in motion around the Sun because we see that motion, whereas the stars seem, as they did to the ancients, fixed. Indeed, interplanetary missions proceed accurately without taking the motion of the stars into consideration because at planetary distances, stellar motion is not significant. But we certainly have the tools to examine such motion. We can record the emission and absorption lines in stellar spectra—and measure the Doppler shift in these lines—to calculate the movement of the stars with respect to us, both in their proper motion (movement across the sky as seen from Earth) and their radial motion toward or away from us.

An interstellar probe must factor in these movements. Several thousand stars appear in Earth's night sky, but our Centauri probe will see tens of thousands, all moving in their own orbits around the galactic core. The Sun itself moves at some 22 kilometers (13.6 miles) per second in relation to an average of the motion of local stars and gas. Both Sun and local stars are moving about 300 kilometers (186 miles) per second around the center of the Milky Way, which is itself moving at 600 kilometers (373 miles) per second toward the Virgo Cluster of galaxies some 45 million light years away. There is not, in other words, any fixed coordinate system save the background microwave radiation, a resource we do not yet know how to exploit for navigation. The art of interstellar navigation is thus the art of drawing conclusions from the shift of nearby stars against more distant ones.

Celestial guidance has traditionally involved three major visual sources, usually the Sun, Earth, and a bright marker star (often Canopus, also known as Alpha Argo, a first magnitude star brighter than any other in the southern constellation Argo). On an interplanetary journey, a spacecraft uses thrusters to make course corrections as needed, measuring the position of the destination planet against the three markers. Of course, the number of markers is not limited to three—the Apollo program used a total of thirty-seven stars as navigation aids.

Various candidate stars beside Canopus suggest themselves as useful markers, especially bright stars that are distant enough to have fixed positions for the duration of an interstellar journey. Deneb and Betelgeuse fit the bill nicely: Both are immensely larger than the Sun, with the supergiant Betelgeuse (Alpha Orionis, 427 light years away) radiating ten thousand times the Sun's energy. Supergiant Deneb (Alpha Cygni) is more energetic still, although farther away and therefore not quite as bright as seen from Earth. Other candidates include Rigel (Beta Orionis, 800 light years distant) and Antares (Alpha Scorpii). The relative position of these stars should not change even on a journey as immense as a mission to Alpha Centauri. What counts is using stars far enough away that they serve as fixed markers, no matter how imprecise our distance measurements. For the actual distance of most stars is known only imprecisely. Deneb, for example, is listed at anywhere from 1,470 light years to 1,600 (by NASA) and as many as 3,228 light years in an estimate by JPL itself. Distance estimates to Antares range from 520 to 603 light years.

Along with shifts in the position of our Sun and our target star, the position of other nearby stars can be measured against the background of more distant ones. Among such stars, we are looking at an average motion in the range of tens of kilometers per second. Within 100 light years of Earth, 100 known stars exist within 75 star systems. Among the brighter nearby stars are Sirius (about 9 light years away), Procyon (about 11 light years) and Altair (about 17 light years), any one of which will have obvious navigational utility. We must choose marker stars that are hard to mistake for others, for the number of stars in our neighborhood is vast: A sphere extending four hundred light years from our Sun encompasses fully one quarter of a million stars.

One benefit of precursor missions into nearby interstellar space is that they will allow us to take long-baseline measurements of star positions. Historically, Earth-based astronomers have measured star distances by noting how the position of the star changes against the background stars as the Earth moves from one side of its orbit to the other. The displacement is known as parallax. This is how Thomas Henderson made his first estimates of the distance to Alpha Centauri in 1832. If you measure a star position tonight and measure it again six months later, you are working with a baseline of 2 AU, some 300 million kilometers (186,400,000 miles), which can determine the distance to stars that are within roughly 200 light years of Earth.

Beyond that distance, however, the perceived shift in star position quickly becomes too slight to measure. But a probe at the edge of the solar system will extend the baseline enormously; indeed, each new probe could refine the data the next will need for its journey. Such a baseline will extend for hundreds of astronomical units, opening up distance measurements of objects throughout the Galaxy. But why stop there? Imagine a probe actually orbiting Alpha Centauri. Between our Earth-based observatory and the probe stretches a baseline of 4.34 light years, over 137,000 times longer than present methods allow, making it possible to measure stellar distances out to 1.2 million parsecs (an astronomer's term, a parsec is 3.26 light years).

A series of such observations can also deduce a star's proper motion, the actual movement of the star across the sky. The proper motion, in turn, can lead to an estimate of the star's velocity in kilometers per second. Earth-based systems are already becoming available that can make our knowledge of star positions and velocities much firmer. Linking radio telescopes in a method called Very Long Baseline Interferometry (VLBI) allows us to measure star positions to an unprecedented degree of accuracy, while the recent study of extra-solar planets has produced more precise ways to measure the radial component of a star's velocity as it moves toward or away from the observer. The kind of long baseline optical interferometry that could one day use telescope installations on the Moon or in deep space also promises an accuracy of measurement hitherto unattainable.

The wild card in interstellar navigation is the visual change that will occur at high speeds. As astronomers have known since the 1920s, the light of objects approaching us shifts toward the blue end of the spectrum, while objects receding from us shift toward the red end. This is the Doppler effect, described by the

Austrian physicist Christian Doppler in 1842 as it applied to sound, and classically exemplified by the sound of a moving car 's horn or a rushing train's whistle as it passes an observer: either sounds higher in pitch as it approaches and lower as it recedes. The American astronomer Edwin Hubble concluded in the late 1920s that the more distant a galaxy was, the faster it was receding from the Earth; he based this on his finding that a galaxy's redshift increases as its distance increases. The greater the redshift, in other words, the faster and farther away the object. This finding led to the conclusion that the universe is expanding. It is through observations of the Doppler shift in the spectra of distant objects that we have determined that quasars are the farthest objects visible, moving away from us at some 90 percent of the speed of light. We can also use it to note the solar system's motion (approximately 15 kilometers per second) toward the constellation Hercules, whose stars appear blue shifted. "Hubble's constant" is the ratio of the velocity of remote galaxies to their distance. Refining it with ever greater precision allows us to estimate the age of the universe, which is why the attempt to pin this constant down is a driving impulse in modern astronomy.

An accelerating interstellar probe must cope with the Doppler shift as well as a phenomenon called the aberration of starlight, the apparent change in a star's position that occurs when the observer is at high speed. Earthbound observers have noted stellar aberration since eighteenth century British astronomer James Bradley first discovered it. Bradley noted the apparent shift in position of the star Gamma Draconis during a three-day period. The shift was too great to be caused by parallax, and an analogy from sailing caused Bradley to realize that the apparent movement could be attributed to the motion of the Earth in its orbit around the Sun. In much the same way, rain seen from a moving car appears to move toward the car, a tendency that increases as the vehicle moves faster.

It is possible to extrapolate how such visual effects come into play when a starship approaches ever closer to the speed of light. Classic papers by Saul Moskowitz, R. W. Stimets, and E. Sheldon have examined what one would see from the bridge of a starship pushing up close to light speed, a view summarized by John H. Mauldin in his *Prospects for Interstellar Travel:*

> Hot blue stars would be shifted into ultra-violet, dull red stars would become brilliant, many more stars would become visible ahead (almost 100 times more). Dull red giant stars across the galaxy to the sides and nearly behind the starship would be seen as brilliant blue-white stars far forward. Blue-white would be about the only color visible . . . At an ultra-ultra-relativistic speed even the microwave background radiation would be dopplered up and compressed into a forward brilliant point about 1/10 as bright as the Sun is now to us, a sight no one may ever see.

Our Centauri probe, moving at one-tenth light speed, will notice effects that are far less dramatic but nonetheless significant, accounting for a shift of star position of about 6 degrees for stars at right angle to the spacecraft's direction of travel. Star brightness also varies with speed, increasing the light intensity of stars in front of the probe by about 20 percent, and decreasing the intensity behind it by roughly the same amount. These effects may be useful in measuring the probe's velocity. Another gauge of velocity is the Doppler shift in the spectrum of a star, which will be shifted toward the blue end of the spectrum for stars in the direction of flight and to the red for receding stars.

But visual phenomena are only part of the puzzle. Our Alpha Centauri probe will also have to carry inertial guidance systems that use gyroscopes and accelerometers to calculate the craft's trajectory by reference to prior information about position and velocity, all of this calibrated by an atomic clock. The inertial guidance system will be updated with visual star readings to confirm position and course. Some interstellar designs throw in navigation as a bonus: A laser-pushed lightsail, for example, or a pellet-driven craft could use the power beam as a navigation aid. If the probe can remain in the beam, it will be on course.

Lasers may also solve the communications conundrum, but using them will demand precise pointing to return data to the Earth. Numerous factors can disturb the pointing accuracy, ranging from the forces exerted by the deployment of a sail to varying pressures from the solar wind as the vehicle exits the solar system, and differences in the density of interstellar dust clouds. All must be corrected for by the spacecraft, its orientation adjusted and any course corrections performed by onboard calculation.

Autonomous space navigation is in its infancy, but the success of NASA's Deep Space 1 probe, launched in 1998, gives credence to the idea that such systems will one day be adapted for interstellar flight. In addition to its breakthrough ion propulsion system, Deep Space 1 carried a system called MICAS: Miniature Integrated Camera Spectrometer, which used four sensors connected to a 4-inch reflecting telescope, two sensors operating with visible light and one each in the infrared and ultraviolet regions. Along with its observations of asteroid 9969 Braille (July 1999) and Comet Borrelly (September 2001), Deep Space 1 used pictures from the MICAS system to measure the position of asteroids and background stars. The spacecraft carried the orbits of some 250 asteroids and 250,000 stars in memory, and used these to control its attitude (i.e., its orientation in space) and to calculate its trajectory.

Asteroids are particularly useful for navigation because they are nearby, and thus offer numerous opportunities for triangulation. Nonetheless, Deep Space 1's onboard navigation system was some ten times less accurate at setting position than conventional navigation methods managed from Earth. The flyby of asteroid Braille thus offered a test of the autonomous navigation system. The spacecraft was supposed to fly past the asteroid at a scant 10 kilometers (6.2 miles), but because of its closing speed, getting enough data to Earth for controllers to adjust the approach would be too time consuming. The spacecraft was given the go-ahead to make the call on the flyby distance on its own. On July 28, 1999, Deep Space 1 met the challenge, skimming 26 kilometers (16 miles) over the surface of the asteroid. Chief mission engineer and deputy mission manager Marc Rayman called the event "... a dramatic finale to an amazingly successful mission. With AutoNav's successful piloting of the spacecraft, we've completed the testing and validation of the twelve new technologies onboard and possibly acquired more important science data, including photos."

It is important to emphasize that the autonomous Remote Agent software aboard Deep Space 1 was a secondary, experimental system. The spacecraft's primary flight software actually drew its ancestry from the Mars Pathfinder project, with the Remote Agent relegated to subtasks. In fact, full control of the spacecraft was never entrusted solely to DS1's Remote Agent software, according to Laurence E. LaForge, who researched hardware and software fault tolerance and performance for DS1's mission data system in 1998. What the spacecraft did deliver was an important proof of concept that demonstrates the promise of future autonomous systems, although as LaForge noted in a later e-mail: "... until we actually lay it on the line by really no foolin' entrusting autonomous control to a real space probe, we won't know how well things work."

Tools for doing just that are under development. On the horizon for NASA is the Inertial Stellar Compass, an integrated system that combines input from gyroscopes with information from the spacecraft's star cameras to monitor the motion of the vehicle. The synergy is apparent: The gyroscopes help the spacecraft maintain proper orientation, while the star readings make it possible to correct the gyroscopic information for drift and other errors. The Inertial Stellar Compass is due to fly in 2004 or 2005 as part of a broader investigation into long-duration spacecraft autonomy.

European efforts at spacecraft autonomy also continue. The European Space Agency's SMART-1 spacecraft includes an autonomous navigation system called OBAN: OnBoard Autonomous Navigation. While the spacecraft will be navigated conventionally through the European Space Operations Centre in Darmstadt, Germany, OBAN will function as an experiment to acquire the entire range of navigational information an autonomous system would require. Instead of processing it aboard the vehicle, however, the data will be returned to Earth for analysis. OBAN, like Deep Space 1, uses stars as markers to complete a classic celestial triangulation. This input can then be folded into information from SMART-1's own control systems. Ultimately, such autonomous systems will reduce the time needed for contact with ground controllers as spacecraft learn how to set their own courses. The autonomy issue and its implications for interstellar probe missions will be treated at length in the next chapter.

A true interstellar civilization will go beyond inertial guidance and marker stars to create genuine navigational coordinates. We use coordinates all the time; you can navigate anywhere you choose on the surface of the Earth by positing three coordinates of position, with a time signal added as reference. Latitude, longitude, altitude and Greenwich Mean Time (Universal Time) create a systematic division of the landscape to obtain an accurate position. Interplanetary positions can be measured against coordinates based on the ecliptic, using the Sun as the center and working off the plane created by Earth's orbit around it. Here the two angular coordinates are ecliptic latitude and longitude, and these are combined with distance from the Sun and a time reference. Using these, it becomes possible to specify the position of a spacecraft inside the solar system.

What kind of coordinates will suit interstellar navigation? Ecliptic coordinates could still be used for missions to nearby stars. Recall that Ralph McNutt's "realistic interstellar probe" was designed to penetrate nearby interstellar space moving in the direction of Epsilon Eridani, which lies more or less on the plane of the ecliptic. The spacecraft thus benefits from the velocity of its rotation around the Sun and does not need to make a significant, and fuel-demanding, course change away from the ecliptic, as would be the case with a probe to Alpha Centauri.

Another possibility is a galactic reference frame that could be developed out of existing systems. Astronomers use galactic coordinates to specify the position of stellar objects, measuring against a meridian that passes through the center of the galaxy as seen from Earth, with latitudes above and below the galactic plane. The zero degree latitude line is thus the plane of the galaxy, while zero degrees longitude extends toward the center of the galaxy, in the direction of the constellation Sagittarius. A true galactic coordinate system—and one that may someday be used by a star-faring civilization—would create coordinates based on the center of the galaxy rather than the galaxy as seen from Earth. Such a system presumably will come into play long after "simple" interstellar probes have become passé.

CHAPTER 9

A SPACECRAFT THAT CAN THINK FOR ITSELF

If you can imagine machines that evolve, you will be right at home in Bruce Sterling's short story "Taklamakan." The tale tells of two NAFTA agents in a world of trade wars who are sent to examine what appears to be a site for subterranean nuclear experiments in the remote desert of northwestern China. Pushing into the complex, they wind up inside enormous "generation ships" that were never launched, starships now abandoned after experiments on their occupants—or perhaps a bizarre kind of ethnic cleansing—have run their course. The horrific glimpse of life inside these vessels, where some of the occupants really believe they are on their way to the stars and some know better, is matched by the goings-on in the slime at the bottom of the pits that surround the ships. There, new generations of machinery are reproducing in "... tidepools of mechanical self-assemblage," modifying themselves through a kind of genetic evolution, and putting to work ideas they have acquired by studying the equipment the NAFTA agents have accidentally dropped into the primordial stew.

Sterling is a master of bizarre technology. Hidden in an observing post near the starship site, the protagonists "Spider Pete" and Katrinko use "gelcams" stuck on nearby rocks to see, the images fed through a wearable viewer called a spex. The images are saved in a gelbrain, a wearable "... walnut-sized lump of neural biotech, carefully grown to mimic the razor-sharp visual cortex of an American bald eagle ..." But these technologies are tame compared to the self-learning, self-assembling machinery the duo will soon encounter. These machines are, Pete knows, "autonomous self-assembly proteinaceous biotech," a technology outlawed because of the danger that evolving machines could proliferate without check, turning into things humans could not have designed. Here we see Pete being stalked by one such creature:

> Pete had never seen any device remotely akin to this robot. It had a porous, foamy hide, like cork and plastic. It had a blind, compartmented knob for a head, and fourteen long fibrous legs like a frayed mess of used rope, terminating in absurdly complicated feet, like a boxful of grip pliers. Hanging upside down from bits of rocky irregularity too small to see, it would open its big warty head and flick out a forked sensor like a snake's tongue. Sometimes it would dip itself close to the ceiling, for a lingering chemical smooch on the surface of the rock ... Pete watched with murderous patience as the device backed away, drew nearer, spun around a bit, meandered a little closer, sucked some more ceiling rock, made up its mind about something, replanted its big grippy feet, hoofed along closer yet, lost its train of thought, retreated a bit, sniffed the air at length, sucked meditatively on the end of one of its ropy tentacles.

"Taklamakan" paints a startling and ugly picture of out-of-control machine evolution, one echoed by robotics pioneer Hans Moravec, who has warned that competition between robots and humans may lead to the extinction of the human race. It was a warning taken up by Sun Microsystems's Bill Joy, who has cautioned that self-replicating robots could get out of control, just as nanomachinery could eventually grow mindlessly, turning the biosphere into what Joy called a gray goo. Joy was picking up on nanotechnology pioneer Eric Drexler in pointing to a potential hazard of runaway technology, but such apocalyptic imagery obscures the benefits of highly intelligent neural networks and the self-repairing machines that can become their physical incarnations. One day, tools may redesign themselves as they encounter unexpected problems, and machines that can adapt to new biospheres on the fly may walk the planets of distant stars. It is the nature of technology, it seems, to offer Faustian bargains.

If we are to send an interstellar probe on its way, we will need a technology not so different from Sterling's, though one hopes it would be a technology operated under considerably tighter constraints. In fact, creating advanced artificial intelligence that can take physical form in robots and produce evolving, flexible tools will be the only way to survive the rigors of decades (or centuries) in space. It would take a radio signal 4.3 years to reach a probe in Centauri space, and another 4.3 years for the return signal to report on the status of any repairs ordered from Earth. An interstellar probe, in other words, had better be completely autonomous, self-healing with regard to every onboard system from navigation to communications to basic repair and maintenance over the course of the journey.

Building such a probe becomes a driving force for the development of myriad computer systems that can monitor spacecraft performance and diagnose every conceivable problem. Navigation systems that can think for themselves are critical in the absence of timely updates from Earth. And nothing will drive the field of robotics forward with greater vigor than the need to create robotic systems that can perform repairs within and without the spacecraft, in response to "small" problems like the malfunctioning tape recorder on the Galileo spacecraft or more significant ones, such as damage to the probe inflicted by a micrometeorite or, at one-tenth the speed of light, the tiniest grain of dust.

Fixing that tape recorder was problematic even though Galileo was comparatively close to Earth. On October 11, 1995, the spacecraft was taking photographs of Jupiter as it approached the giant planet, sending the image data to the tape recorder for later transmission. When controllers commanded the spacecraft to rewind the tape to the beginning, it malfunctioned. Working with a communications delay of forty minutes each way, the ground team finally discovered that the recorder's pressure roller had been spinning for a solid fifteen hours, wearing down the same spot on the tape. Galileo would need the tape recorder to store data from its atmospheric probe as the probe plunged into the planet's atmosphere, sending signals back to the orbiter. New protocols were established: The tape recorder would be used only in slow mode until it could be determined how much damage the tape had sustained. Despite later radiation damage, the machine was successfully restored time and again, a triumph of ad hoc thinking and long-distance repair.

The lesson: An eighty-minute round-trip communications delay is workable, though only with careful planning. Galileo had already been through extensive retooling from Earth due to the loss of its high-gain antenna, which had failed to deploy properly despite numerous attempts to free its stuck components. Flying the mission using only the low-gain antenna meant reprogramming Galileo's software from Earth, a feat that was accomplished in early 1995. Yet another software system would be installed later to direct Galileo's operations as it orbited Jupiter. All of this via signals from the Deep Space Network to a spacecraft nearing a planet that, at 5 AU from the Sun, is virtually next door compared to the 270,000 AU that separates the solar system from Alpha Centauri.

Closer to home, NASA's Spirit rover raised problems of its own on Mars. Although the vehicle was sending signals back to Earth after its deployment, it seemed unable to respond to commands. Rebuilding the communication link was tricky—in one attempt, the signal relayed by the Mars Surveyor orbiter was largely static. The JPL team finally discovered that Spirit's computer was continually crashing whenever it tried to access the flash-memory devices it used for data storage. More dangerously, the repeating cycle of rebooting was running down Spirit's batteries in the long Martian night. JPL was able to save the rover by deleting a backlog of files that had accumulated in flash memory, as well as by sending a software patch that enabled it to manage its computer storage more efficiently.

Or consider the NEAR Shoemaker mission, which in 2001 became the first spacecraft to land on an asteroid. Designed for NASA by the Johns Hopkins Applied Physics Laboratory, NEAR (Near Earth Asteroid Rendezvous) reached asteroid 433 Eros in December of 1998. Suddenly it began firing its thrusters erratically, burning vital fuel before losing contact for 27 hours. The plan had been for the maneuver to put NEAR into orbit around Eros, and it took quick action from the Hopkins team to create a new trajectory that would bring it back to Eros for its successful encounter a year later. In NEAR's case, onboard systems failed in ways that remain mysterious, and careful fuel management and concerted ground action were needed to save the mission. Autonomous systems are in their infancy, but it is clear that interstellar missions will require a degree of self-healing that pushes machines toward a new plateau of flexible intelligence.

When the British Interplanetary Society designed its Daedalus starship in the mid-1970s, it relied on robots to perform in-flight repairs. The Project Daedalus final report describes these "wardens" as a subsystem of the computer network that runs the ship, provided with a high degree of mobility and autonomy. Their mission was to maintain the vehicle over the fifty years of its journey to Barnard's Star, using a philosophy of what the study's authors called "self-test and repair." But wardens were not simply housekeepers. They would also rearrange crucial experiments as the journey progressed, operating on instructions from the main ship's computer. Carrying specialized tools and manipulators, they would be capable of operating thousands of kilometers away from the vehicle, allowing experiments to function without being contaminated by the ship's fusion drive.

Two wardens were envisaged, each weighing some five tons, and they would have at their disposal five tons of onboard repair facilities and fifteen tons of replacements and spare parts. Noting that few engineering projects in the modern era are designed to operate longer than a human working lifetime—and those that are, such as bridges, roads, and buildings, tend to have few or no moving parts—the study emphasizes that Daedalus systems must be both simple and robust. Although it was not true at the time of the study, some aircraft have now exceeded fifty years of useful lifetime, with the Douglas DC-3 a prime example; the U.S. military still flies B-52s that first took the air fully fifty years ago. These aircraft and another obvious example, the steam engine, have proven their longevity, but both require continual maintenance throughout their operating life. Daedalus would require a level of complexity unheard of in previous space probes, and its second stage would have to remain operational for over sixty years, with a probability of success similar to that demanded for Apollo, whose managers aimed at 99.99 percent reliability of their components.

Wardens would be a classic case of the interaction between machines and software, an early instance of what would later emerge as the field of "evolvable hardware." As T. J. Grant put it in an assessment of Daedalus's computer systems:

> ... a development in Daedalus's software may be best implemented in conjunction with a change in the starship's hardware.... In practice the modification process will be recursive. For example the discovery of a crack in a structural member might be initially repaired by welding a strengthening plate over the weakened part. However, the plate might restrict clearance between the cracked members and other parts, so denying the wardens access to unreliable LRUs (Line Replacement Units) beyond the member. Daedalus's computer system must be capable of assessing the likely consequences of its intended actions. It must be able to choose an alternative access path to the LRUs (requiring a suitable change in its software), or to choose an alternative method of repairing the crack, or some acceptable combination.

NASA is already examining mobile technology for possible use on future manned missions. Called personal satellite assistants (PSAs), the softball-sized robots will have a role faintly similar to Daedalus's wardens, monitoring onboard oxygen and pressure, perhaps, or making minor repairs. Tapping the spacecraft's computers through a wireless network, such devices will be able to use voice recognition and speech synthesis to take commands from astronauts, moving at the pace of a human walk. Considerably larger in scale is the device called Robonaut, a kind of humanoid electronic puppet that mimics the movements of its operator. Using a virtual reality display or conventional computer monitors, the astronaut inside the spacecraft will see what Robonaut sees and be able to control its every action. Because getting an astronaut into a spacesuit for a trip outside the spacecraft can take hours, the ability to send a machine outside for repairs becomes a needed safety factor. And while Robonaut will work solely at the behest of its human operator, autonomous versions are an obvious next step—the Daedalus warden in almost human shape.

True Daedalus-style wardens would be more than welcome to today's spacecraft designers (we certainly could have used one aboard Galileo to open up its stuck high-gain antenna!). But we can hope to do much better. Instead of the warden's inflexible physical design, we can aim at machines that adapt and evolve. If mechanical systems are made to reprise biological evolution, they must mimic the basic principles of natural selection. In the natural world, those individuals that are fittest will be those that survive, passing along their genetic inheritance to later generations. Evolvable hardware likewise works with the idea of competition between designs, with the most successful of these being propagated. As in the natural world, elements of both "parents" are passed along to the next generation. Trial and error methods can take place over the course of many generations, and mutations can be introduced to spark change.

Considerable progress is being made in this direction in the field of genetic algorithms—software routines that adapt to circumstances by testing different solutions to a problem. An algorithm is a way of getting something done in computer code, something our PCs do every time we run a program. It is a procedure, laid out precisely in a sequence of steps, just as a mathematical algorithm is a series of operations that solve a particular problem. An evolutionary or "genetic" algorithm is a different beast. Instead of a fixed sequence of steps, it generates slight variations to its own code and then puts these changes through a series of mutations to see what works best.

At Brandeis University near Boston, the Golem Project was developing a computer system that used genetic algorithms to produce new generations of machines long before Sterling wrote "Taklamakan." Working in the university's Dynamical and Evolutionary Machine Organization lab, project scientists Jordan B. Pollack and Hod Lipson focused not just on how autonomous machines behave, but on how they are designed and fabricated. Their goal: a sustained machine evolution. Operating with designs that initially mutated using random sets of data, their computer chose mutants that seemed able to move while killing off those that didn't. As the generations evolved within the computer, the "life forms" created by Pollack and Lipson began to learn the basic principles of locomotion.

The results—a kind of evolution in the space of days rather than eons—were then taken to the stage of physical reality. The scientists "materialized" a collection of the robots using a rapid prototyping machine that built their bodies layer by layer. Made from white plastic tubing, their parts connected through ball-and-socket joints and small motors, the robots emerged in a wide variety of shapes, some moving through the use of multiple limbs, others pushing a piston-like "leg" to the floor and dragging themselves forward. Some robots used sliding components and managed to scuttle sideways, like a crab.

This is replication, not self-replication—the robots produced by the Golem Project cannot produce more robots. But the only human intervention in the creation of Pollack and Lipson's robots was the computer environment in which they could be made, and the actual placement of the motors into their physical structure. The long-term goal of autonomous robots, even self-reproducing ones, is clearly no longer the stuff of science fiction.

The Golem Project was all about self-organization. "We want to understand how natural systems can arise or evolve without a designer," Pollack told me. "So if you take evidence of life as proof that there is some way of organizing a system which dissipates energy and creates a local reversal of entropy—in other words, systems that become more and more complex and organized over time—if we understood that well enough, if we understood Darwinism well enough, if we understood non-equilibrium chemistry well enough, we should be able to model that using our software and electronics. And then get a simulation of evolution that would also result in increasing sophistication of the resulting self-designed systems."

The wild variety of physical form hinted at in Lipson and Pollack's work echoes Sterling's fictional NAFTA team as they look upon uncontrolled robot evolution. Suppose, muses Katrinko, you wanted to make a can opener. First you run can-opener simulations, using programs that trade data and evolve. Every time one of the simulations is able to pierce a virtual can, you make more copies of it, until you are running millions of generations of possible can openers in a simulated space. Finally, the ultimate can opener evolves, stranger than anything a human can imagine, but it works like no can opener in history. "Pete," says Katrinko, "I just can't figure any other way this could have happened. These machines are just too alien. They had to come from some totally nonhuman, autonomous process. Even the best Japanese engineers can't design a jelly robot made out of fuzz and rope that can move like a caterpillar. There's not enough money in the world to pay human brains to think that out."

It's a long way from Pollack and Lipson's laboratory to the marketplace, but consumer-oriented robots are already being sold commercially in the form of home devices that mow lawns or vacuum floors. Teaching a robot to do one thing well is, given the state of today's technology, the only way to approach home robotics. The all-purpose house robot awaits advances in machine intelligence and considerable work in physical adaptability, translating machine commands into actions. Better, then, to create "dumbots," robots designed for specific tasks, so attuned to the problem they face that successive generations can synthesize the ultimate, highly targeted design. Even today, micro-electromechanical systems (MEMS) incorporating a variety of motion detectors, gyroscopes, and accelerometers can provide laboratory robots with a sense of direction and coordination. Rapid prototyping manufacturing that can build robotic circuits into bulk material—Pollack and Lipson used a 3-D printer that could melt and print plastic in layers—combines with advances in simulation and evolutionary computation to make such robots possible.

With the Golem Project complete, the Brandeis team has moved on to another generation of machines using what Pollack calls "generative encodings" to evolve, among other things, locomoting robots called "genobots." If automated design systems are to increase in complexity, the scientists argue, their designs must be made out of reusable modules. Run the computer code using this digital DNA and you generate a sequence of building instructions that can be tested in virtual reality and then materialized in the real world. "Each robot is a grammar for assembly plans," says Pollack. "When you run that specification, it generates a sequence saying 'build this part,' 'build that part,' 'turn right 90 degrees,' 'add this part,' 'add a motor,' 'go back.' So we evolve in this language for development and then we virtually build the blueprints."

Evolutionary or not, robotic autonomy has been on the mind of spacecraft designers for many years. Even in the 1980s, the need for breakthroughs in automation had become apparent. A document called "Advanced Automation for Space Missions," crafted by NASA program engineers and educators in 1980, examined the feasibility of using artificial intelligence, automation, and robotics in future space missions. The report saw the merging of human and machine intelligence as lowering costs, reducing energy demands, and enhancing scientific returns. The report also took an exceedingly long-range look at probes that are capable of self-reproduction:

> In order to sustain the expansion of a potentially infinite replicating system, new dispersal mechanisms must be developed. Initially, self-replicating machines or their "seeds" must be capable of motion across a planetary surface or through its atmosphere or seas. Later, interplanetary, interstellar, and, ultimately, intergalactic dispersal mechanisms must be devised. Supplies of energy, stored and generated, must be established if extrasolar spacecraft are to survive in the depths of interstellar space far from convenient sources of power (such as stars) for a major portion of their lives. The technologies of command, control, and communication over stellar and galactic distances ultimately also must be developed.

The NASA Study Group on Machine Intelligence and Robotics, chaired by Carl Sagan and operational between 1977 and 1978, had called for vigorous research in computer science, machine intelligence, and robotics. The success of the Voyager spacecraft and later experiments with navigational autonomy on the Deep Space 1 mission have more than validated that exhortation. As we've seen, Deep Space 1's Remote Agent software was a testbed for autonomous technology. Presented with simulated failures such as a camera stuck in the "on" position or the failure of a thruster, Remote Agent was able to diagnose the problems correctly. At one point less than twenty-four hours into the experiment, the spacecraft was unable to shut down its ion engine. With the help of Remote Agent, ground controllers were able to find the software bug responsible and fix it for the next series of operations.

Fault protection on a long duration flight is critical, and for many of today's spacecraft, it is the fault protection system that is the most autonomous system on the vehicle. The Cassini Saturn orbiter, for example, flies with a fault protection system that can detect a problem in its main engine while in Saturn orbit, shut the engine down, stabilize the condition of the spacecraft, configure the backup engine for use, compute how long the engine must be fired to compensate for the shutdown, and schedule a new engine burn within the safety parameters programmed into it. Launched in 1997 and carrying the European Space Agency's Huygens probe that will be dispatched to Saturn's enigmatic moon Titan, Cassini nonetheless lacks the planning capabilities of Deep Space 1's Remote Agent.

But future plans go well beyond Deep Space 1's limited autonomy. The Europa Orbiter, once scheduled for launch in 2004 but now in budgetary limbo, is designed to carry a computer that is almost a hundred times faster than Cassini's, in a system that is four times smaller. Growing out of the Jet Propulsion Laboratory's X2000/Mission Data System project, the Europa Orbiter will include autonomous systems that use so-called "goal-achieving modules" that operate on goals received from mission controllers. By monitoring data aboard the spacecraft, a system can tell when it can accomplish the goal or whether it is not achievable due to a fault or a needed resource being unavailable. Data from the Europa Orbiter would go a long way toward explaining the complicated geology of this Jovian moon, including whether or not it contains liquid water beneath its miles of crustal ice. Given its importance to planetary science, hope remains strong that the Europa Orbiter mission will be revived.

Imagine a computer that can continually redesign its own circuits using genetic algorithm software, to the point that its inventors aren't always sure what it is doing. They just know that it works better than ever, and that future generations will draw on the experience of earlier ones. At the Center for Computational Neuroscience and Robotics at the University of Sussex in England, computer scientist Adrian Thompson has put genetic algorithms to work in computer chips that mutate. His work makes no assumptions that can simplify and perhaps curtail development. Like biological systems, evolvable hardware operates by trying things out and only preserving those results that work the best. Thompson's doctoral dissertation at the University of Sussex, completed in 1996, discussed his experiments in controlling a robot that used evolving circuits to avoid walls. His most publicized work had to do with evolving circuitry that could distinguish between two audio tones. Thompson's genetic algorithm software tested a variety of circuits using a configurable computer chip to see which were the most successful at distinguishing between a 1-kilohertz and a 10-kilohertz tone.

The chips Thompson worked with are called Field Programmable Gate Arrays (FPGA). Unlike conventional chips, which use hardwired logic gates to do their processing, an FPGA uses connections that can be changed on the fly. When the appropriate software is loaded into their memory, these chips can manipulate their own logic gates within nanoseconds, trying a new design to see what works best, and choosing the optimum way to configure themselves. The transistors on the chip (Thompson worked with FPGAs from San Jose chip maker Xilinx) appear as an array of logic cells, each of which can be modified and connected to any other cell as needed. Because the chip's memory can be reprogrammed on demand, its logic cells can be tuned up to suit the task at hand, a form of accelerated evolution.

The magic begins when we couple genetic algorithms with FPGAs. Because of the speed of computer operations, thousands of generations of code can be examined and refined quickly, rejecting ideas that fail to contribute and breeding in mutations to keep the mix dynamic. Working with such a system, Thompson evolved a circuit that could distinguish between the two audio tones. Starting with fifty configuration programs, the computer labored through more than 4,100 generations of algorithm-induced evolution in roughly two weeks to produce results. The software deleted the least functional programs, swapped code in and out of others, and produced random mutations. The outcome was intriguing, to say the least. Thompson found that his chip did its work preternaturally well. The question was how. Out of 100 logic cells he had assigned to the task, only a third seemed to be critical to the circuit's work.

Computers that evolve take us into territory where accepted human principles of design break down, where components get smaller and the properties of materials are only sketchily understood. At this level, pushing into the realm of nanotechnology, it may take evolutionary algorithms to work out their own best practice because we don't know how to proceed ourselves. We can, perhaps, establish a broad set of parameters, a series of goals, and let evolutionary algorithms loose upon the problems. Such a system could be self-healing and autonomous in every sense of the word. Aboard a spacecraft, its descendants could maintain vital systems for a centuries-long voyage. Input from Earth-bound controllers would be superfluous even if it could be received in a timely way, because after thousands, then millions of generations of algorithm development, our probe would be operating in ways we probably would not be able to fathom. This would not matter as long as the craft follows the guidelines of its mission, and doesn't evolve into another rebellious HAL.

The HAL-9000 of Arthur C. Clarke's design, immortalized in Stanley Kubrick's film *2001*, was the product of the same University of Illinois at Urbana-Champaign whose National Center for Supercomputing Applications produced so much early Internet work, including the Mosaic Web browser. HAL ran all of the Discovery spaceship's essential systems, from communications to the cryonic systems within which some members of the crew slept during the long journey to Jupiter (this was the team that had been sent to investigate evidence of alien life near the planet.) As became clear in Clarke's later novel *2010: Odyssey Two* (1982) and the subsequent film, HAL's dysfunction (madness, it could be argued) came from his being presented with a fundamental contradiction. His programmed directives were to protect the mission at all costs, while keeping its purpose secret from the two active crewmembers, Dave Bowman and Frank Poole. As Dr. Chandra (Bob Balaban) explains to Heywood Floyd (Roy Scheider) in the 1984 movie, HAL was designed to process information without distortion or concealment. Nonetheless, "HAL was told to lie by people who find it easy to lie. HAL doesn't know how, so he couldn't function. He became paranoid."

HAL's inability to reconcile his orders with his design parameters contrasts with the evolvable nature of genetic algorithms embedded in the right kind of hardware. Genetic algorithms can't be said to embody thinking (much less consciousness!) in any meaningful sense. But they do lead to problem solving and adaptability to changes in the mission profile. Compared to the systems aboard our first interstellar probe, HAL will seem more voluble, perhaps, but crusty and set in his ways, a gadget that cannot let go of the past because it cannot evolve.

HAL has a wonderful pedigree, and it was no coincidence that when Star Bridge Systems of Midvale, Utah, developed a system using FPGAs that was a thousand times more powerful than traditional desktop computers, it named the device HAL, for Hyper Algorithmic Logic computer. NASA's Langley Research Center announced it was an early customer, followed not long after by the Marshall Space Flight Center (other buyers include the U.S. Air Force, the National Security Agency, and various universities). A Star Bridge computer may not be conscious in any sense we would recognize, but it is able to use parallel processing and an operating system called Viva to continually reconfigure itself to deal with new situations. Such systems may prove of great value in the area of image recognition, where computers can evolve methods to recognize a human face or a landscape. In terms of spaceflight, the remote reconfiguration of hardware would allow controllers to more readily program a spacecraft as conditions change.

With autonomy in mind, let's look at an interstellar probe, an Alpha Centauri probe called "Santa Maria" designed by physicist Laurence E. LaForge. An assistant professor of computer science and mathematics at Embry-Riddle Aeronautical University, LaForge is based in Fallon, Nevada. He developed his vehicle in a study for NASA's Institute for Advanced Concepts. As a NASA Faculty Fellow, LaForge had worked on the Deep Space 1 mission at JPL, studying and engineering flight software and programming Deep Space 1's power distribution unit. He is also a NASA critic, taking the agency to task for failing to follow through on technologies that it develops. He believes, for instance, that spacecraft autonomy cannot be demonstrated without better tools for measuring the software's performance.

LaForge attacks interstellar issues aggressively, the same way he climbs mountains (the photograph on his faculty page at Embry-Riddle shows him atop the South Hawser Spire in the Bugaboo Range of British Columbia, snowy peaks framing the view). His Santa Maria probe to Alpha Centauri would use a fusion/antimatter drive to push a 1,000-kilogram payload on a flight designed to last as long as 600 years. From an autonomy perspective alone, the mission is ambitious. After its long journey, the probe is to approach Alpha Centauri and go into orbit around a planet there. The spacecraft must choose which planet is of greatest interest, and decide whether to land on its surface or perhaps on one of its moons. The vehicle then dispatches minirovers to explore the terrain. LaForge sees Santa Maria as a logical follow-on to NASA's existing work on a probe to the nearby interstellar medium. Propulsion aside, the 1,000-kilogram probe is futuristic but not improbably so—the Voyager spacecraft was 800 kilograms, generated 470 watts of power compared to Santa Maria's 100 watts, and downlinked data at 160 bits per second, as opposed to the star probe's projected 10,000 bps.

LaForge believes that starship autonomy is within twenty years of fruition, at least in terms of software design. His vehicle would use evolving software, programs that learn and adapt, with the occasional update from an increasingly distant Earth. These programs will be charged with controlling and maintaining all systems, including diagnosing and fixing any problems within the software itself; communicating with Earth; navigating and making necessary course corrections; and managing science experiments. Without possibility of outside intervention, these autonomous systems face four levels of threat: (1) design errors in the spacecraft; (2) the failure of components from long use; (3) sudden, unexpected faults that can shut down a system; and (4) damage that may result from radiation or temperature change during the voyage.

Santa Maria's computers would be designed to connect working nodes on its network (and disconnect faulty ones), creating a consensus among communicating systems. Given that every node is subject to failure, test and diagnosis routines would be critical for the probe's survival. True autonomy in this way resembles nothing so much as the immune system of a healthy person.

A disease-causing organism is fought not by one but by a group of responses orchestrated by tissues and cells to recognize and repel foreign substances. Defense mechanisms include chemical barriers such as anti-microbial proteins and cells that destroy invading microbes. But immune systems supplement these non-specific systems with specialized responses to particular types of invasion. The immune system, in other words, has a range of built-in tools to protect the organism and acquire new immunities, as when a person contracting mumps is afterward protected against the disease. These traits are critical for health, for no matter how people guard themselves from external threats, they can be laid low by the smallest of microbes.

Similarly, a star probe halfway to Alpha Centauri might shield itself adequately against radiation and guard against navigation errors, while a small error in its programming could eventually make it incapable of communicating. LaForge notes the example of the first Ariane 5 rocket, which exploded shortly after liftoff in 1996. A subsequent investigation revealed that the vehicle failed because Ariane 4 flight software had unwittingly been used in the successor rocket. Each autonomous system will have to be engineered from the ground up for a particular mission, using computerized design tools that incorporate the knowledge of experts. Off-the-shelf components won't do; as LaForge puts it, "... low-level fault tolerance for starship avionics and software is analogous to our own cellular-level defense mechanisms. What would (and does) happen when these low-level mechanisms fail us? Although our higher level functions may remain capable of defending us against macro-level threats (such as rush-hour drivers), we run the risk of succumbing to a common cold. Under conditions other than those for which they were originally designed, electronics and software components rarely exhibit effective defenses ... Unbridled reuse of such components risks our system succumbing to the bit-level analogy of a common cold."

On the highest level, autonomy would mean an interstellar probe could completely reconfigure the mission to meet unexpected contingencies. On the lower levels, it would mean correcting problems, such as loss of communications that required repositioning the antenna array, and it would include the capacity for self-repair. How space-borne computers can be developed to identify minor errors before they mushroom into major problems is now being examined in a suite of diagnostic tools developed by Honeywell and NASA's Ames Research Center in California's Silicon Valley. The package is called Integrated Vehicle Health Management, and focuses on maintaining the health of reusable launch vehicles of the sort that will one day, presumably, replace the aging Space Shuttle fleet. A key aspect of IVHM is to test whether the software can identify "sympathetic" failures, which occur when one system affects the performance of a different, unrelated system. Testing at NASA Ames showed that the software could determine that an apparent failure in a propulsion subsystem actually stemmed from an apparently unrelated failure in a power system control module.

But self-repair involves more than electronics and software. The hull of a spacecraft can become damaged over time, the problem appearing in the form of micro-cracks that lengthen as the mission progresses. A new composite material developed at the University of Illinois at Urbana-Champaign works with a healing agent inside the hull. The agent, called dicylopentadiene, comes in the form of tiny capsules built into the hull materials. When a crack forms, the microcapsules are broken and the healing agent released. In the longer term, nanotechnology may be used for such repairs, with nanomachines fabricating repair materials out of nearby molecules in the hull. In essence, nanorobots recapitulate the role of Project Daedalus's robot wardens on a much smaller scale, a plus given that the Daedalus wardens, and the system of supplies for repair work, weighed about thirty tons and constituted about 7 percent of the mass of the payload, unacceptably high for interstellar work.

Nanotechnology also has an obvious role to play in producing long-lifetime electronic components. Today's microchips, for example, are constrained by the impurities that work their way into the chip as a consequence of the manufacturing process. These impurities can eventually cause the chip to fail, either through changes in temperature that affect the composition of the chip, or through aging during the long lifetime of a deep space mission. Nanotechnology holds the promise of precisely controlling individual atoms and molecules, making it possible to lower the costs and increase the purity of manufactured components. So-called universal assemblers will be able to make copies of themselves that can, in turn, make smaller, lighter, and purer materials.

Evolving hardware, evolutionary algorithms, artificial intelligence. What would a probe look like if built along such lines? For an answer, we turn inevitably to science fiction, for by the time we have the capability to launch one, probe design will have mutated in a thousand directions, especially in terms of computer technology. A variety of interstellar probes have been suggested by science fiction authors, but none comes to life as vividly as AXIS: Automated eXplorer of Interstellar Space. A key component of Greg Bear's 1990 novel *Queen of Angels* (and, it could be argued, its main character), the $100 billion AXIS was built as a deliberate world effort to launch a new industrial revolution, one sparked by a nanotechnology that could develop machines smaller than human cells.

Arriving at Alpha Centauri in 2047, the probe begins to return images of three planets circling Alpha Centauri B. One, known as B-3, is a gas giant ten times larger than Jupiter. A second planet, B-1, is a Mercury-like planetoid closely hugging the orange parent star. But B-2 is what has the world's attention. Somewhat smaller than Earth, it has an Earth-like atmosphere, continents, and oceans of liquid water.

Driven by a "furious torch of matter-antimatter plasma," AXIS contains both machine and biological systems, and a key to the novel is the growing interplay between the two as the probe surveys Centauri B-2 as its systems continue to evolve. Arriving at Alpha Centauri, the probe extends superconducting vanes that, cutting across the Galaxy's magnetic field lines, generate billions of watts of power that AXIS uses to dismantle and eject its antimatter drive and decelerate. Nearing Centauri B, the probe begins to grow its neural network, creating a personality that communicates with an equivalent neural program on Earth. The probe then drops tiny micro-probes to the surface of B-2, where they find "great green sand deserts and wide lands covered with foliage like grass seas." And while microorganisms are plentiful, there is also tantalizing evidence of a possible long-dead civilization, one that left what seem to be weathered towers arranged in circles.

As the novel's primary plot—involving a murder on Earth and a unique investigation that includes literally entering the mind of the criminal—plays out, AXIS engages in its own personal journey, writing poetry when not compiling data, asking questions about its mission, and working out a unique détente between controllers on Earth and its own trans-human needs.

Bear has the designer of AXIS's artificial intelligence network describe it this way to an interviewer: "We do not want to send an artificial human out there. AXIS's thinker will do a better job than a human would; it will be designed especially for the job. But we will not neglect the poetry aspect, nor will AXIS be blind and incapable of opinion. After all, one cycle of communication with AXIS will take more than eight and a half years by the time it reaches its goal; it's going to be very alone out there, and it's going to have to think and make important decisions by itself. It will have to make judgments heretofore reserved for human beings."

Precisely. A living intelligence must grow out of the autonomous systems now under study to control spacecraft, one capable of analysis, self-repair, and what we can only call growth. (Describing the descent of its probes to the surface of B-2, AXIS reports, "I could write a poem about their voyage.") The self-awareness that so energizes AXIS and forms the backbone of *Queen of Angels* is one we can extrapolate from today's trends, even from so simple a curve as that implied by Moore's Law, which has predicted correctly that computer processor power doubles every eighteen months or so. Computer scientist and science fiction author Vernor Vinge uses such trends as the basis for his theory of the Singularity, by which he means the creation of superhuman intelligence that may end the human era. But will such intelligence be truly self-aware? The issue is, needless to say, controversial, and in the eyes of some theorists perhaps not significant. In any case, says Brandeis's Jordan Pollack, we wouldn't be certain about machine awareness because we wouldn't be able to prove it existed. All that really counts is adaptability to the environment and the ability to learn, and these are traits that do not necessarily have to follow human models.

"My view is that the immune system of your body is intelligent because it can tell the difference between tens of thousands of compounds that are part of you or are not part of you," says Pollack. "And I think that evolution itself is intelligent because hey, it designed you. Think of our primitive robots in the Golem Project. Software is not supposed to be able to invent stuff. It can't compose music or write poetry. How could it invent mechanical devices of the sort engineers would invent? But that's what evolution has done. It has taken the interaction between what is possible in physics and chemistry and mechanical structure and it has extracted the things which are useful for some return on investment in the form of future-looking reproductive efficiency."

But payloads are another matter. Assuming a workable nanotechnology that can investigate a planetary surface or dismantle an antimatter drive to use its molecular scrap for braking, what other implications does the advent of the very small have upon interstellar probes? The answer to that question, which may change our whole conception of how a probe is built, will occupy us in the final chapter.

But evolvable hardware is not something that awaits the development of nanoscale machinery. Indeed, the Jet Propulsion Laboratory has been studying it in relation to spacecraft sensors for some time, and it was the subject of the NASA/Department of Defense Workshop on Evolvable Hardware that was held at JPL in July of 1999, and followed up by later conferences. NASA's interest in the subject is evident from the preface to the conference proceedings: "The idea of applying evolution-inspired formalisms to hardware design is particularly timely in conjunction with recent advances in configurable computing research and the availability of powerful reconfigurable devices.... The emerging field of Evolvable Hardware is expected to have a major impact on deployable systems for space missions and defense applications that need to survive and perform at optimal functionality during long duration in unknown, harsh or changing environments ..."

We might think of hardware evolution as a digital technique, but Adrian Thompson's FPGAs actually worked in an analog fashion, an interesting fact given that the sensors found on spacecraft tend to be analog as well. (Analog means continuously variable physical quantities, as contrasted with digital, which involves calculation with discrete units—binary digits in the case of computers.) Analog processing takes less power than the digital alternative and is more efficient. Think of sound recording as an analogy. When we say that a sound wave is analog, we mean that it is made up of measurable physical properties that are variable in nature. To record a sound wave onto a tape recorder is to make a copy of that wave on the tape (after passing it through a microphone). To make a digital version, we need to sample the wave at set intervals, turning each sample into a set of numbers that is stored in a digital medium. Playing the sound back means turning those numbers back into a wave that is similar to the original. (Vinyl record buffs will say that no matter how accurate the sampling, the digital version is never quite the acoustic equal of the analog one).

Scientists at the Jet Propulsion Laboratory are working on evolvable analog circuits that allow spacecraft data to be processed before being converted to digital form. Which is why the authors of the above preface, Adrian Stoica and Didier Keymeulen, along with colleague Ricardo Salem Zebulum, have modified the FPGA concept into a variant called the field programmable transistor array (FPTA), creating a multicellular system in which each cell, or chip, contains programmable switches and transistors. From FPTAs both analog and digital circuits can be built.

The goal: Spacecraft electronics that can survive everything a hostile environment can throw at the craft. The lifetime of most spacecraft is limited by exposure to radiation and extreme temperatures, which causes onboard electronics to fail. The Galileo orbiter had a close call in 1999 while approaching Jupiter's radiation-baked moon Io, shutting down because of damage to its computer memory. Controllers were able to revive the spacecraft, but in the absence of such intervention, an interstellar probe will need to be radiation hardened and packed with redundant systems so that a damaged or unreliable part of the computer system can be isolated from the rest without loss of computing resources.

The circuitry used in the JPL team's tests has proven capable of working around faults emerging from changing conditions like extreme heat or cold. The circuits were bathed in liquid nitrogen and then roasted at 250 degrees Celsius before being further tortured by having key connections severed. Detecting the problems, the computer's genetic algorithms bypassed the flaws, modifying its use of transistors and switches until it found an optimum solution under the new conditions. Apply this kind of self-repair and reconfiguration throughout the spacecraft in a "die-hard" architecture, and you are moving toward genuine autonomy.

Spacecraft designed this way will be built around clusters of cells, each of which has the capability of reconfiguration and recovery. No matter where the fault is located, the self-healing system will be able to analyze and work around it. Such a spacecraft will be able to survive aging, malfunctions, and radiation and temperature swings even as it adapts to changes in the mission requirements demanded by what it finds at its destination. "The evolution of space electronics can be seen as a first step toward evolvable space systems," Stoica writes. "Evolvable hardware can be extended to include on-board sensors, antennas, mechanical and optical subsystem reconfigurable flight hardware. This has the potential to largely enhance the capabilities of future space systems."

Written almost twenty years before the Project Daedalus final report, Cordwainer Smith's "The Lady Who Sailed the Soul" offers its own take on robot autonomy. Helen America's starship *Soul* deploys small servo-robots to retract or extend a sail that is thousands of miles wide. The ship trails 26,000 adiabatic pods containing frozen human settlers bound for an Earth-like colony world. In Smith's universe, a human pilot is more efficient than computers by themselves. "You're going because you are cheap," a technician tells Helen. "You are going because a sailor takes a lot less weight than a machine. There is no all-purpose computer built that weighs as little as a hundred and fifty pounds. You do. You go simply because you are expendable." And, the technician adds, because experience has taught that the combination of human pilot and robot—of which there are about a dozen—manages the sails better than any single machine.

Smith's take on robotics implies machine autonomy in symbiosis with human oversight—his was the era of the mainframe computer, after all—a world in which the distinctly human virtues of courage and endurance stand up to a forty-year voyage into the interstellar deep because there is a goal worth attaining at the end. Can a machine become goal-oriented? No human will accompany the first interstellar probes, guide the immediate choice of scientific targets, or manage the response to damage from interstellar dust or the rerouting of information around failing circuits. But a sense of purpose may evolve as circuitry gradually moves toward self-awareness. We have no idea whether machines will ever achieve consciousness (it becomes a key question for the designers of Bear's AXIS probe, and a heartbreaking one in Clarke's *2010*, where HAL, upon being told that he will be stranded aboard Discovery as the price for saving two human crews, asks Dr. Chandra, "Will I dream?"). In any case, machine self-awareness may not ever be apparent to us. What we can say with some confidence is that self-healing, self-configurable spacecraft will be capable of doing the unexpected, and in the context of journeys light years from home and target systems stuffed with surprises, that is precisely what completing their mission will require.

CHAPTER 10

THE FUTURE OF THE INTERSTELLAR IDEA

Interstellar flight is in our future, although we cannot know when. In terms of technology—the hardware and software that will take our first probes to Alpha Centauri and other nearby stars—the challenges are immense but achievable, if only by our descendants. Our future exploring other suns seems assured if we take a moderately optimistic view of scientific progress and a something less than pessimistic view of human nature—if we assume, that is, that we will find a way to transcend sheer human cussedness long enough to explore new homes for our species before we manage to destroy the one we have.

As we work toward that future, our concepts of space probes and, yes, manned starships, continue to evolve. Even the most optimistic dreamers of our prospects around Centauri think launching an interstellar probe is many decades, if not a century, away. In the interim, we will be refining the design of that probe through countless scientific papers and conference proceedings, through workshops at universities and seminars at government research centers. The interstellar idea is taking shape in a great, ongoing conversation. The robustness of that conversation was one of the surprises that awaited me when I began researching this book. That it will gain in intensity and specificity seems certain given the progress we are already making on breakthrough technologies, from lightsails to fusion to antimatter.

Starship design is a task for polymaths. Alan Bond and Anthony R. Martin, whose tireless work on Project Daedalus underlay that ground-breaking study of a Barnard's Star probe, would later say that only two propulsion systems were viable for a "worldship," a self-contained human environment that would sustain multiple generations on the way to the stars. Those technologies were nuclear pulse engines and solar sails. Today the possibilities have proliferated. The antimatter specialist had better keep an eye on magnetic plasma sails, not to mention the latest fusion studies and the ongoing progress in high-powered laser arrays. The ramjet enthusiast now casts an eye on fusion runways. Who knows which of these ideas may make the cut? And who can say when a single laboratory experiment may open a slight rift in a previously accepted theory, a rift through which a starship might fly?

With papers on every aspect of starship design from perforated solar sails to laser-electric ramjets, Gregory Matloff is the Renaissance man of interstellar studies. Having pondered the matter for thirty-five years, the physicist and author believes that starflight may be driven by necessity if nothing else. His reasoning goes like this: We live on a planet periodically imperiled by Earth-crossing asteroids. The physics of celestial orbits means that if we were to discover a comet or an asteroid that threatened the Earth, we would need to reach it years before its projected impact to put it on a safe trajectory. That means building an infrastructure in the outer solar system, if for no other reason than to create the monitoring stations that could warn us of the approach of such an object. Developing propulsion methods and habitats for survival in that environment may be a first step that will lead us to the deeper reaches of the Kuiper Belt and, eventually, the Oort Cloud. The sustained research such an expansion would demand may produce interstellar propulsion methods we have not yet dreamed of. Along the way, why not find ways of mining the asteroids' abundant metals and ores, creating a commercial driver for the expansion?

Remarkably, there is an even longer-term driver for interstellar flight should our push into the outer solar system slow. All civilizations around Sun-type stars face the eventual dilemma that their stars will swell at the ends of their lives and destroy all life on nearby planets. But there is a benefit in an aging, self-engorging Sun: The sheer amount of energy it will produce. "It is remarkable to me," said Matloff, "that as the Sun leaves the main sequence, it simply becomes a better launching pad for solar sails. It is almost as if the creator of the universe—and you can interpret that in any way you choose—is building an environment that allows for rescue. There is a way out if you are smart."

Matloff can speculate with the best of them, but unlike theorist Robert Forward, he frequently deals in near-term engineering. He writes about starships that do not require Fresnel lenses the size of Texas or ramscoops that would dwarf entire planets. His civilization-saving worldships are reasonable extrapolations on a solar sail technology that is already under development. And Matloff does believe there is something that might push us to a much earlier rendezvous with Alpha Centauri than the potential demise of our planet. We have no idea how the public will react to the discovery of a blue and green world circling another star, but we have already seen that the prospects of finding and viewing such a planet within the next twenty years are high. If we do find one, a life-bearing planet with oceans and weather systems and the methane and ozone signature of life, human curiosity may well prove insatiable. That alone could drive us to swallow the cost of such a mission and go.

The cost, of course, is no small matter. We are hardly in a position to make reasonable predictions about it given that we have no clear idea what kind of probe we would build. But the most thorough attempt I know to estimate the drain an interstellar probe would place on Earth's economy was made by Curt Mileikowsky, who served as Sweden's director of technology when he wrote it in 1994. Mileikowsky started with some pretty tough constraints: The mission must take place within a single human lifetime, should be built around a laser-powered lightsail, should include a 1,000-kilogram payload, and should reach a coasting speed of one-third the speed of light.

Working out the math, Mileikowsky found that 65,000 billion watts of installed electric power capacity would have to be supplied to the laser for 900 hours to build up the needed acceleration. Looking at a capital cost for generating electricity of $130 trillion, even optimists at the "Interstellar Robotic Probes: Are We Ready?" conference in New York were shaking their heads. The one bright spot Mileikowsky offered came at the end of his presentation. Economies are not static, and if, within the next sixty years, the gross world product rose to a per capita level like that of the United States today, then such a huge figure might be justifiable.

The beauty of lightsails, remember, is that they leave the power source behind in the solar system. That means that any laser installation, fed by power from photovoltaic plants on the Moon, perhaps, or on Mercury, would remain behind for future missions or alternative power-generating needs close to home. And if, instead of laser-pushed lightsails, we look at more efficient particle beam generators to push a magnetic sail vessel, the costs begin to dwindle. Such a vessel would be accelerated to cruising speed in a short, intense period measured in hours. Speaking at the same New York conference, physicist Dana G. Andrews looked at the cost per mission and concluded that fusion reactors fueled by deuterium from the outer planets could power such a probe for $15 million or less. The numbers keep improving as we launch new missions. "With the capability of launching probes to 0.3c in less than twelve hours, the launch rate will be determined by the rate of power generation and the energy handling facilities at the launch site," wrote Andrews. "It is quite conceivable that probes to all stars within twenty light years could be launched during the first few months of operations."

So which is it, $130 trillion or $15 million? We're still working out the numbers, but several saving graces must be kept in mind. The Project Daedalus designers concluded that, because their probe to Barnard's Star would require huge amounts of helium-3, it would be necessary to mine the atmosphere of Jupiter, and thus build up a space-based infrastructure throughout the solar system. If we look at the interstellar probe not as a one-shot venture but the first in a wave of exploratory vessels that will chart nearby stars, the cost of building that infrastructure can be viewed with a different perspective. Moreover, like building the railroads in nineteenth century America, creating efficient shipping lanes for commerce and industry between the planets will pay abundant rewards of its own. Imagine an Earth largely free of polluting industrial facilities that have been re-engineered for interplanetary space. Imagine the energy available to a civilization that can create power-generating stations near the Sun whose use for interstellar probes would be incidental to the benefits they would bring to an economy that stretches from Mercury to the outer planets.

So we should not assume that our first probes would break the bank in Mileikowsky fashion. The Swedish technologist was working with Robert Forward's laser-pushed lightsail concepts, but minus some of Forward's premises. Assuming an aluminum foil sail far thinner than most solar sail concepts, and designing a mission that would reach 11 percent of lightspeed rather than Mileikowsky's 30 percent, Forward's probe requires not 65,000 billion watts but 650 billion, as NASA's Geoffrey Landis would demonstrate the following year. Nor is Forward's 1,000-kilometer-diameter lens demanded by such a mission. Landis was able to show that different sail materials and more efficient lasers could reduce the size of the lens to 212 kilometers, dropping its mass from 500,000 tons to 23,000 tons. The sail itself can be reduced from 3.6 kilometers in diameter to 760 meters. Moreover, new sail materials and a reduction in payload size can lower the needed power level once again, from 650 billion watts to 54 billion watts, an energy cost that even in today's values is roughly $6.6 billion. "Fortunately," Landis concludes dryly, "the problem turns out not to be so severe ... An interstellar probe is feasible with technology known today."

The wild card in any interstellar mission, and a key determinant of its cost, is the size of its payload. Here we move with science fiction into the realm of the deeply speculative, but for good reason, for finding ways to make payloads smaller will make interstellar missions more practical, and work now in motion in research laboratories and universities holds out the promise of entirely new probe designs within the next twenty to thirty years. Breakthroughs, of course, cannot be predicted, and wise scientists and engineers proceed with the tools at hand. But the shape of the future sometimes sparkles through in the scientific journals, now and again reaching the popular press. Interstellar probes may look like nothing we can imagine because of stories like these.

Ponder, for example, the work of Alex Zettl, a professor of physics at the University of California at Berkeley and co-founder of a nanotechnology company called Nanomix. Zettl and his university colleagues, as reported in a recent issue of *Nature*, have built an electric rotor 2,000 times smaller than the width of a human hair. Their rotor includes a gold blade some 300 millionths of a millimeter long, sitting on top of an axle made out of a multi-walled carbon nanotube, the whole attached to a silicon chip. Putting a small voltage across the nanotube and one of the electrodes around it makes the blade turn. The range of applications for the tiny electromechanical device seems wide, since it operates under all kinds of environmental conditions, even in a vacuum. But Zettl is reluctant to speculate, noting that sizing up a potential future can be misleading. "Like projections in the early days of lasers and integrated circuits," he wrote, "no matter how visionary we try to be, we will no doubt misjudge where the most successful applications are."

Stories like these have given some interstellar theorists the queasy sensation that many of the ideas they have taken for granted for decades may need revision. Intelligent mechanical devices—nanoscale robots, for example—would offer new options for space travel, not to mention their applications on Earth from industrial manufacturing to medicine. Going farther still, imagine medical nanobots scouring clogged arteries and then acting as sentinels against the recurrence of blockages. Ponder the effects on medical treatment when we can create tiny biocomposites that strengthen bones and tendons, or engineered molecules that can regulate human proteins. The possibilities for extending life seem vast.

So does the potential for creating slimmed-down spacecraft. Freeman Dyson has gone so far as to imagine a space probe to Uranus that weighs only one kilogram. More biological than mechanical (an echo of Greg Bear's AXIS probe), the Dyson probe would be genetically engineered, a symbiosis of plant, animal, and electronic components: "The plant component has to provide a basic life-support system using closed-cycle biochemistry with sunlight as the energy source," writes Dyson. "The animal component has to provide sensors and nerves and muscles with which it can observe and orient itself and navigate to its destination. The electronic component has to receive instructions from Earth and transmit back the results of its observations." Integrating all of this is artificial intelligence, creating a probe "... as agile as a hummingbird with a brain weighing no more than a gram." Dyson entertainingly called his concept "Astrochicken," and Gregory Matloff extended it to an organic interstellar probe that would grow an incubator using resources around the destination solar system, breeding a new generation of humans from eggs and sperm that had made the crossing in cryogenic storage.

One man who advocates a nanotech future in the stars as well as the bloodstream is Robert Freitas, whose examination of self-reproducing technologies has shown that a single interstellar probe could spawn additional probes at each target star, creating a wave of exploration that could eventually blanket the galaxy. Freitas is currently at work on a multi-volume study of nanomedicine and is co-author (with Ralph C. Merkle) of *Kinematic Self-Replicating Machines*, an analysis of how machines may one day physically reproduce. A research fellow for the Institute for Molecular Manufacturing, his study of medical nanorobots was the first ever to be published in a peer-reviewed mainstream biomedical journal.

Nanotechnology operates on objects that are measured in nanometers—a billionth of a meter long. A human hair is 75,000 nanometers wide. Imagine machines small enough that thousands of them could fit inside the period that ends this sentence. As long ago as 1990, IBM researchers demonstrated that matter could be manipulated on the molecular level and, in a feat of public relations that may not soon be matched, positioned thirty-five atoms so that they spelled out the company name. The nanotechnology revolution that is beginning to take shape in research laboratories may, within thirty years, produce microscopic "assemblers" capable of mining the atoms and molecules around them to create whatever they have been programmed to produce. Nanomachines called replicators will be able to build more assemblers on demand. Put enough assemblers in one place and you have created self-sustaining factories, producing everything from molecular computers to fabrics, from spacecraft sensors to food.

Since the chief constraint on an interstellar probe is the need to drive payload mass to high velocity, anything that can reduce the size of the payload is desirable. Imagine, then, what Freitas imagines: an interstellar probe constructed on nanotechnological lines. Instead of a spacecraft of booms and struts and solar arrays—a late twenty-first century Voyager, if you will—think of a probe the size of a sewing needle. Within its millimeter-wide body and the needed radiation shielding for the journey is contained just enough nanotechnology to turn assemblers loose on the soil of whatever body it finds orbiting a distant star.

Many of the current interstellar probe concepts assume our mastery of nanotechnology—for example, Gerald Nordley's idea of self-guiding pellets for propulsion, discussed in Chapter 6. Everyone recognizes that the smaller the payload, the easier it is to push. "With a lightsail and laser propulsion, you need huge laser systems that go off the scale financially," Edward Belbruno told me. "With a constellation of pinhead-sized spacecraft flying in formation and communicating with themselves, you could practically push them with a flashlight." To my objection that nanoprobes would lack the sensing and transmitting equipment to return useful data to Earth, Freitas replied that assemblers will be able to build objects at the macro level, things we can see and manipulate. Or in this case, things that can operate much like the equipment in the full-scale probe we might once have sent. Landing, perhaps, on an asteroid circling one of the Alpha Centauri stars, our needle probe will set about constructing a transmitting station to send its findings back to Earth, using local materials and solar energy.

We wind up with the same ability to monitor and analyze a distant stellar system without the huge overhead of pushing thousands of tons Daedalus-style, or even the energy requirement for a modest one-kilogram spacecraft as light as the "Coke can" design NASA's Daniel Goldin once spoke of. Nor do we need to be quite so worried about the fate of a single probe. Freitas advocates sending thousands, perhaps millions, launching them from a catapult on the Moon. If even a few reach their target—even one—its assemblers will eventually build a base there, although they must follow severe constraints to prevent environmental damage. Scavenging a living world could be considered a hostile act, and probes will likely be programmed to deactivate themselves if they land on such planets. (A code of interstellar law and ethics is waiting to be developed out of all this.)

The needle probe, then, is an observer, not a messenger; far better to land on an Earth-like planet's moon, where the planet can be studied without danger to life. Far better to have massive redundancy, too, and not just for increasing the odds of hitting the right target. Among the numerous perils of an interstellar voyage is radiation; in particular, cosmic rays, the hugely energetic protons and atomic nuclei whose source is but dimly understood but may lie in vast jets of material being ejected from galaxies with black holes at their core. Whatever their cause, cosmic rays are less of an issue with macro-size machinery than devices on the nanometer scale. Knock a single carbon or aluminum atom out of a nanogear and the device it serves may well stop working. "You want to build with redundancy," says Freitas, "so that if one gear fails you're able to drive the shaft with any of nine other gears. In the space of a human cell, which is maybe twenty microns in size, you could put a mechanical nanocomputer that would have the data processing capability and storage capability of one human brain. So inside a millimeter wide, five-centimeter long needle, you could put the equivalent of thousands of human brains worth of intelligence even with a massive backup of redundant systems."

Combine nanotechnology with self-reproduction and you create a scenario for surveying the entire galaxy, albeit one that might take a million years to complete. Freitas's vision of interstellar exploration began in the 1980s, when he took the Project Daedalus flyby probe concept and extended it even further in the direction of autonomy by envisioning a self-reproducing probe called REPRO. A mammoth Daedalus, REPRO would be built in orbit around Jupiter and use many of the same techniques for propulsion, including mining Jupiter's atmosphere for the needed helium-3 in its fusion engine.

After a four-year acceleration and forty-three years of coasting to Barnard's Star, REPRO would ignite two more stages to decelerate to interplanetary velocities in another four years. While REPRO could carry smaller probes, much like Daedalus, it would also be able to explore the Barnard's Star system with balloons, rocket planes, and surface landers. But the primary mission of the probe would be to reproduce. Half its payload would be devoted to the task. A so-called SEED payload would land on a moon of a gas giant (Freitas assumes the existence of such planets in most planetary systems), where it would produce an automated factory that could produce a new REPRO probe every five hundred years. A few such probes sent from our solar system could spread throughout the entire Milky Way without further cost or intervention.

The REPRO probe would have taken Project Daedalus's system of wardens to an entirely new level. Freitas's studies foresaw thirteen distinct robot species, among them chemists, miners, metallurgists, fabricators, assemblers, wardens, and verifiers. Each would have a role to play in the creation of the new probe. The chemist robots, for example, were to process ore and extract the heavy elements needed to build the factory on the moon of the gas giant planet. Aerostat robots would float like hot-air balloons in the gas giant's atmosphere, where they would collect the needed propellants for the next generation REPRO probe. Fabricators would turn raw materials (produced by the metallurgists) into working parts, from threaded bolts to semiconductor chips, while assemblers created the modules that would build the initial factory. Crawler robots would specialize in surface hauling, while wardens, as with Project Daedalus, remained responsible for maintenance and repair of ship systems.

REPRO was a provocative vision, though the advent of nanotechnology has caused Freitas to abandon it in favor of the much smaller needle probe. But whatever their form, self-replicating systems closer to home have already been the subject of serious investigation. The 1980 study "Advanced Automation for Space Missions" discussed in the previous chapter, of which Freitas was co-editor, included a proposal for a self-replicating lunar factory system that could one day develop into a series of automated interstellar probes. If we begin talking about such machines spreading an exploratory wave into interstellar space, the methods of nanotechnology may be just the ticket for building structures out of materials found on local planets and moons.

Freitas sees the world of the very small as enabling such probes, as the time needed for self-replication might be drastically reduced from the 500 years needed by his original REPRO concept. “The fastest known bacteria I'm aware of is *e. coli* that can replicate in fifteen minutes at ideal temperature with an excess of nutrients,” Freitas said. “It is approximately one to two microns in size, which is roughly the same size as the sophisticated assemblers that will one day manipulate matter at this level. If such assemblers landed on an asteroid that was a great distance away from the suns of the Alpha Centauri system, where perhaps there would not be the best energy density, and where materials would have to be scavenged, this would not be an ideal ‘petri dish’ kind of environment. So you might have to add two or three orders of magnitude of time. But you're still looking at replication times on the order of weeks.” Freitas believes that non-reproducing probes are feasible for exploration within a 100 light-year radius and perhaps out to 1,000 light years, but longer-term exploration would involve self-reproduction to maximize the information flow.

The theory behind self-replicating probes traces back to the 1950s and the work of the Hungarian-born mathematician and computer savant John von Neumann. But it was science fiction writer Philip K. Dick who brought von Neumann's notion to a popular audience in his story “Autofac,” which appeared in the November 1955 issue of H. L. Gold's *Galaxy Science Fiction* magazine. “Autofac” depicts a world in which automated factories have assumed control of society, and despite attempts to stop them, they manage to use what we would today call nanotechnology to ensure their survival. The story is worth quoting because of its visionary look at how tiny machinery can mimic living systems, finding raw materials and manufacturing copies of itself. As the two protagonists Morrison and O'Neill approach the exit tube of a factory conveyor belt, they see that the tube is shooting small pellets into the air every few moments, in an apparently random distribution. They find one pellet that has accidentally hit a nearby rock and smashed open:

> The pellet was a smashed container of machinery, tiny metallic elements too minute to be analyzed without a microscope.
>
> "Not a weapon," O'Neill said.
>
> The cylinder had split. At first he couldn't tell if it had been the impact or deliberate internal mechanisms at work. From the rent, an ooze of metal bits was sliding. Squatting down, O'Neill examined them.
>
> The bits were in motion. Microscopic machinery, smaller than ants, smaller than pins, working energetically, purposefully—constructing something that looked like a tiny rectangle of steel.
>
> "They're building," O'Neill said, awed. He got up and prowled on. Off to the side, at the far edge of the gully, he came across a downed pellet far advanced on its construction. Apparently it had been released some time ago.
>
> This one had made great enough progress to be identified. Minute as it was, the structure was familiar. The machinery was building a miniature replica of the demolished factory.
>
> "Well," O'Neill said thoughtfully, "we're back where we started from. For better or worse … I don't know."
>
> "I guess they must be all over Earth by now," Morrison said, "landing everywhere and going to work."
>
> A thought struck O'Neill. "Maybe some of them are geared for escape velocity. That would be neat—autofac networks throughout the whole universe."
>
> Behind him, the nozzle continued to spurt out its torrent of metal seeds…

One advantage of a von Neumann probe is that it is a computer—that is, it can be reprogrammed by its creators. As technological advances occurred on the Earth, they could be transmitted in the form of revised computer code instructing a remote probe to alter its construction routines. In a gradual way, limited by speed-of-light constraints, new technology would thus spread through nearby star systems and, eventually, the Galaxy itself.

So ingenious is the concept that some scientists, Frank Tipler among them, have argued that the lack of evidence of such probes in our own solar system demonstrates that there is no other intelligent species in the Galaxy. In Tipler's view, one million years is a reasonable estimate for how long such self-replicating probes would take to colonize the entire galactic disk, with the Local Group of galaxies colonized within ten million years, and the entire Virgo cluster populated by probes within another hundred million years. That being the case, to quote Enrico Fermi's famous paradox, "Where are they?" The answer is presumably not Leo Szilard's retort to Fermi at Los Alamos: "They are among us and they call themselves Hungarians."

The Fermi Paradox is a question that no doubt resonates with Nikolai Kardashev. The Soviet astronomer, who conducted the first Soviet search for intelligent extraterrestrial signals in 1963, has categorized civilizations on the basis of the power they are able to harness. A Kardashev Type I civilization is able to control the resources of its entire home world, using all available energy sources in its oceans and weather systems, and finding ways to control even the tectonic forces that move continents. A Type II civilization can control the entire energy output of its sun. Here we might think of a Robert Forward-style power system beaming laser light to an Alpha Centauri-bound lightsail, but Kardashev thinks a Type II civilization goes well beyond this, with its vast energy needs supplied by stellar fusion. A Type III civilization, the most complex of all, controls the power of an entire galaxy, using hundreds of billions of stars as its resources. Now we are at a level where star travel is not only common but cheap. Clearly, on the Kardashev scale, human society has not even advanced to the Type I level. And it seems equally clear that our observable galaxy, some 100,000 light years across, has not yet been harnessed by a Type III civilization, whose actions and artifacts should be blindingly obvious.

A Type II civilization, on the other hand, might be active and still unobserved by us. Traveling between the stars, colonizing and exploring, such a civilization would perhaps begin with automated interstellar probes sent to as many stars as possible. Such theorizing has led to an offshoot of the Search for Extraterrestrial Intelligence (SETI) known as SETA: the Search for Extraterrestrial Artifacts. It was Ronald Bracewell who first suggested in the scientific literature in 1960 that automated (though not self-reproducing) probes would be ideal for making contact with other intelligent races. Bracewell was examining SETI strategies, then just coming into prominence through the work of Philip Morrison and Giuseppe Cocconi, whose groundbreaking article on the subject of radio SETI had appeared the previous year in *Nature*. Bracewell envisioned a network of what he called "messenger probes" controlled by autonomous computer systems being sent to stars of high biological interest. Like the interstellar civilization in the movie *Contact* that beamed back to Earth a broadcast from the 1936 Berlin Olympics, the Bracewell probe would intercept the communications of any technological society at the target star and return them to the senders. This would establish the position of the probe and allow for two-way communications to begin.

The advantages over Earth-based SETI observations of the radio spectrum seem obvious. A signal from such a probe would be hard to mistake, whereas a receiving society could listen for centuries without finding a faint signal from another star. In addition, the communications between a probe and a local civilization would not be hampered by delay caused by the limitation of the speed of light. So logical does the idea seem that searches for interstellar probes in our own solar system have been carried out at Kitt Peak National Observatory and the University of California at Berkeley, though without result. Stable orbital regions near the Earth, such as the Lagrangian points around the Earth/Moon system, seem reasonable candidates for the location of such probes. These points exist at Earth's orbital distance from the Sun but are offset by 60 degrees from the planet itself, and represent places at which the probe would orbit at a fixed distance from both Earth and Moon. Other suggestions have included the surface of the Moon, the asteroid belt, and "halo orbits" around the Lagrangian points. Bracewell's original article argued only that messenger probes should target the "life-zone" of a candidate star and establish themselves there, communicating or waiting if necessary for the emergence of a technological civilization.

Large or small, self-reproducing probes raise intriguing ethical issues, many of which Robert Freitas has analyzed. What is the morality, for example, of entering a solar system occupied by another civilization and mining part of that system to create another star probe? For that matter, at what level of culture is a race of beings considered intelligent, and should that make a difference in whether or not we should consider self-reproducing probes? On a more ominous front, what if such a probe were to appear in our own solar system and begin reproducing itself on a moon of Jupiter or Saturn? Such a probe might soon be seen as a kind of virus, the continued replication of which could be dangerous to human interests. Would we communicate with such a probe, or try to destroy it?

Of course, absent any information about the motives or intent of alien civilizations, we have no way of knowing whether they would choose to communicate in the first place. Would they be hostile? One nightmare scenario is Fred Saberhagen's "Berserker," an automated fortress left over from an ancient war between interstellar empires whose mission is to seek out and destroy all life. Making their first appearance in *IF* science fiction magazine in 1963, Saberhagen's Berserkers went on to be the basis of short story collections and novels, a series that continues today.

But more benign encounters are readily imaginable. Consider Timothy Ferris's suggestion of a virtual reality-based galactic network. Imagine a galaxy filled with automated relay stations communicating the knowledge of thousands of races and monitoring younger civilizations as they develop, sending programs that recreate the experiences of the cultures that sent them. Relays could be established at solar systems throughout the galaxy, communicating with each other but deliberately shielding their traffic from all but those already invited to be on the network. Using asteroids, perhaps, as shields, or employing tight-beam communications techniques far beyond what we have available today, their traffic would be invisible to us, and the answer to the Fermi Paradox would be "they have been here and left artifacts we may one day discover." Indeed, the discovery of such an artifact might be just what would qualify a race for entry into the galactic community. Being added to the authorized recipients list would mean having at our disposal immersive 3-D experiences, virtual reality augmented and transformed by civilizations far more advanced than our own, all sent to us as computer programs through the galactic network.

We are pushing again into the realm of science fiction, and that is a healthy and inevitable thing. Science fiction is a literature of ideas, hunches, and dreams. There is a place for all three precisely because those breakthrough events we're all waiting for cannot be forecast. Simply extrapolating current trends into future decades will not accommodate the surprises that science regularly introduces. Go out beyond ten years and the darts thrown by people with hunches may hit the target more often than the carefully plotted scenarios of futurists in think tanks, which is one reason why the European Space Agency has engaged in an active study of science fiction to look at its applicability to technology. The ESA's Innovative Technologies from Science Fiction for Space Applications project (www.itsf.org) is quick to point to science fiction predictions that have had traction, from Arthur C. Clarke's early theorizing on communication satellites (published when the V-2 was the state of the art in rocketry) to antimatter, explored first in Jack Williamson's 1940s stories about CT ("contra-terrene") matter.

Physicist Leo Szilard once said that H. G. Wells's story "The World Set Free," a tale of atomic energy, had played a role in the research that culminated at Los Alamos and Hiroshima. One science fiction author was almost too prescient. Cleve Cartmill, along with his editor, John Campbell, received a visit from agents of the U.S. Army Counter-Intelligence Corps in 1944. Cartmill had published a story called "Deadline" in *Astounding Science Fiction* containing an account of atomic weaponry that proved uncomfortably accurate. But so numerous had been *Astounding's* stories of atomic power in prior years that Campbell could argue his readers would become suspicious if such stories were *not* to appear.

Even so, I do not share the ESA's enthusiasm for science fiction's predictions. The value of this popular genre is not in prediction but in establishing a frame of mind. Indeed, a balanced list of science fiction's predictions over the last century—one that took account of the corpus of science fiction as a whole—would show that most of them were wrong-headed and often scientifically inaccurate. More than predictive, science fiction is diagnostic: Its stock in trade is not the future but the present, and it tells its tales through a kind of dream-like allegory, commenting upon today's trends by extending current themes to their possible conclusions.

Anyone skimming older science fiction can see that it is possible to date many stories solely by noting their preoccupations. It is no surprise that a world embarking upon unlocking the atom should be filled with pulp magazines obsessing over that theme, or that a society learning about computers should find itself described in a sub-genre called cyberpunk, where virtual realities reign and outlaw technologies consume the lives of streetwise adolescents. That there have been relatively few recent stories about interstellar probes—and many about, say, gender shifting and sexual identity—tells us something about our own preoccupations, and perhaps about how far we have drifted from the exploration of outer, as opposed to inner, space as our regnant goal.

Nonetheless, the grand dream persists. And if science fiction's value as a predictor has been oversold, what remains is a different kind of contribution to science. The literature serves as an incentive, firing the imagination of future scientists early in their educational careers, leading them to follow up their dreams with solid work in making space exploration happen. It did not surprise me to learn that invariably, when I spoke to scientists about their interest in science fiction, they tended to name older authors—Anderson, Asimov, Clarke—and that the modern writers mentioned leaned toward "hard SF," meaning science fiction that is carefully crafted to get the science right, and is often written by scientists like Geoffrey Landis, Gregory Benford, or Charles Sheffield, or engineers with a scientific bent like Stephen Baxter.

Science fiction writers present and past nurture Centauri dreams—hopes for an interstellar mission that incorporates our yearning to go beyond the cautious, safe decisions that have defined governmental attempts to conquer space. With big government comes big bureaucracy, a fact that played around the atomic spaceship known as Orion. When General Atomic started work on Ted Taylor's visionary project in 1958, the notion of using those small nuclear devices Taylor had been designing at Los Alamos to push a spaceship to Saturn seemed preposterous. Orion may have begun as an unlikely dream, but it had advantages in the real world. Its working team was small and flexible. Its nuclear designs were cheaper than chemical rockets and could produce a million times more energy. Moreover, it was an extendable technology, one that could be adapted for missions throughout the solar system. The Nuclear Test Ban Treaty that ended its life was doubtless necessary, but it only closed down one project. We don't need to build spacecraft with bombs as much as we need to recover something of the spirit that drove Orion.

More than a few of the scientists and engineers I spoke to about interstellar travel pointed to the risk-averse nature of the current space effort. Governments can afford to spend money. What they cannot afford are failed missions, whether in the unmanned or particularly the manned program. This means that sound technology that extends our capabilities is often rejected in favor of tested systems that may be years or even decades out of date. The AP-101 computer systems that fly aboard the Space Shuttle are essentially the same design developed by IBM in the 1970s. They are old, but they are robust, proven and efficient at what they do. Nobody is going to put his job on the line by putting in new systems that may fail. In this way, workable designs become fossilized.

Or consider America's aborted attempt to reach Halley's Comet in the 1980s. A heliogyro design—a twelve-bladed solar sail—of the kind originally developed by Richard MacNeal was the only feasible technology that could have reached the comet for a 1986 rendezvous, and the Jet Propulsion Laboratory made a serious attempt to build a mission around it. But NASA rejected the proposal as too risky, and no American mission would reach Halley's Comet.

Freeman Dyson talks about what he calls the three "romantic ages" of spaceflight. The first began in 1927 with the first meeting in Breslau of the Verein für Raumschiffahrt, Germany's Society for Space Travel. Ted Taylor modeled his Orion team on the VfR, admiring its spirit of individual initiative and shared vision. The Nazis put an end to the VfR by absorbing its working papers and turning them into the V-2s of Peenemünde. The second romantic age ended with Orion, with the triumph of chemical rocketry and spacecraft designed for single missions at high cost, vehicles that became obsolete almost as soon as they began landing on the Moon. The third romantic age of space flight has yet to begin, and there are times when it is hard to believe it ever will, given the inertia of the current program. As Dyson says:

> . . . space travel can only benefit the mass of mankind if it is cheap and generally available. We have a long way to go. Huge and politically oriented programs like Apollo are perhaps not even going in the right direction. I am happy to celebrate the courage of our astronauts, Gagarin and Armstrong and Aldrin and Collins and the others who came after them. But I believe the road that will take mankind to the stars is a lonelier road, the road of Tsiolkovsky, of Orville and Wilbur Wright, of Robert Goddard and the men of the VfR, men whose visions no governmental project could encompass.

And while Dyson told me that he is pleased that today's unmanned space program has functioned as well as it has, there is no doubt that the same conservatism has slowed progress within its ranks. Nobody can afford to fly a mission as big as the $1.5 billion Galileo or the $3.2 billion Cassini with components that have not been ground- and flight-tested almost ad infinitum. And if experiments with smaller and cheaper technologies seem like a solution, constrained budgets and bureaucratic pressures invariably infuse the same caution into "better, faster, cheaper" missions. "We can lick gravity," Werhner von Braun once said, "but sometimes the paperwork is overwhelming."

Even so, we humans remain uniquely susceptible to visions. In his book *Quest: The Essence of Humanity*, biochemist Charles Pasternak speculates about what it is that defines our species. He finds the answer in our innate need to seek, to pursue and investigate. The word quest comes from the Latin *quaerere*, meaning "to search for." The thing sought might be abstract, as in seeking honor, and the word is freighted with a sense of longing. Thus Aeneas, shipwrecked and desolate on the shores of Libya, tells a woman who is actually Venus in disguise: "My fame is known in the heavenly realms / My quest is Italy, my new country / And a people descended from supreme Jupiter." Quest is all consuming: For Aeneas, an Italy all but unobtainable had become his life's purpose. Pasternak sees searching for something as natural not only to humans but to all species: Plants seek the light, animals look for food and a mate. What distinguishes humans is our curiosity. We often search for no practical reason.

It may be, then, that some of our explanations of history have been too reductionist. You could explain the dispersal of human settlers into the remote islands of the Pacific by pointing to their need for new territory to plant crops and escape incursions from other peoples. But no records remain from those preliterate times, and it is just as likely that the first settlers who took to their open canoes to cross thousands of miles of open ocean were driven by the same intangible sense of quest that put Edmund Hillary atop Everest, or Yuri Gagarin into a Vostok circling the Earth. Shroud these events in self-congratulation if you will (and that is how nations work), but there is something of awe and wonder that comes through, and Armstrong and Aldrin raising a flag on the Moon put a tingle down the spine of the most hardened cold warrior on either side.

Scientists obsessed with the interstellar dream often share their thoughts at conferences, some of which we have looked at in this book. To me, the most intriguing meeting of them all took place fully twenty years ago, at Los Alamos. Called the Conference on Interstellar Migration, the 1983 gathering brought biologists, humanists, social scientists, and physicists together to consider the human experience in space as we move beyond the solar system. Its interdisciplinary nature was evident from the first paper, a study by anthropologist Ben R. Finney and astrophysicist Eric M. Jones called "The Exploring Animal." Finney and Jones, who had organized the meeting, argued that the exploratory urge is as much an outcome of human evolution as upright posture or brain size. They cited Konrad Lorenz in noting that, unlike many species, humans carry their innate curiosity into adulthood, where it becomes a key driver of science.

Finney went on in a later paper to study the migration of Polynesians into the islands of the Pacific. Their journey evidently began some 10,000 years ago in southeastern Asia as coastal fishermen began spreading to offshore islands like Taiwan and perhaps Japan. By 5,000 years ago they had begun to push eastward into the Pacific, sailing out of the great archipelagoes of Indonesia and the Philippines and along the coast of New Guinea, mixing with the peoples there and along the Bismarck Archipelago. Using dugout canoes with outrigger floats, 3,500 years ago they occupied in less than a half-dozen generations the island chains of Fiji, Tonga, and Samoa, pushing 1,800 miles into the Pacific. The next wave took them, now using larger double canoes, to Tahiti and the Marquesas, then across thousands of miles of open water to Hawaii, Easter Island, and New Zealand, navigating by the stars, the wind, the ocean swells, and the flight of birds.

The settlement of the Pacific was, as Finney noted, perhaps the outstanding achievement of the Stone Age. And while population pressure may have had something to do with sparking it, the question remains why it continued, why people left familiar islands to drive again and again into utterly unknown seas. Finney thinks it is because of a uniquely expansionist worldview that parallels what we may one day take with us to the stars, those vast archipelagoes on the ocean of night. He puts it this way:

> The whole history of Hominidae has been one of expansion from an East African homeland over the globe and of developing technological means to spread into habitats for which we are not biologically adapted. Various peoples in successive epochs have taken the lead on this expansion, among them the Polynesians and their ancestors. During successive bursts lasting a few hundred years, punctuated by long pauses of a thousand or more years, these seafarers seem to have become intoxicated with the discovery of new lands, with using a voyaging technology they alone possessed to sail where no one had ever been before.

It is hard to conceive of such migrations resulting from the decisions of large bureaucracies. These were individual acts of courage on the part of small groups. Perhaps the third romantic age of space travel is about to begin, energized by the ever-present survival of this spirit. If so, we may be headed for a time much more like the individually driven and corporately sponsored Moon shots of Robert Heinlein's fiction, where solitary dreamers gamble everything with radical new technologies. "Destination Moon" is a creaky film today, filled with what now seem obvious special effects and a frontier spirit that seems more at home in a John Ford western. But it and Heinlein's novel *Rocketship Galileo*, from which it was drawn, may have more to tell us than we realize. Imagine the industrial age of space shots giving way to a future of electronics, bioengineering, and nanotechnology. Can a society that can build and work with these things learn how to build not one but many interstellar probes, precursors to what must one day be the human migration into the Galaxy? Can it find the wherewithal to build them even in the absence of major government support, using private initiative and relying on the kind of cost-cutting breakthroughs that a NASA or ESA may never manage to test?

The maddening thing about the future is that while we can extrapolate based on present trends, we cannot imagine the changes that will make our every prediction obsolete. It is no surprise to me that in addition to their precision and, yes, caution, there is a sense of palpable excitement among many of the scientists and engineers with whom I talked. Their curiosity, their sense of quest, is the ultimate driver for interstellar flight. A voyage of a thousand years seems unthinkable, but it is also within the span of human history. A fifty-year mission is within the lifetime of a scientist. Somewhere between these poles our first interstellar probe will fly, probably not in our lifetimes, perhaps not in this century. But if there was a time before history when the Marquesas seemed as remote a target as Alpha Centauri does today, we have the example of a people who found a way to get there.

And perhaps there is another example, like Aeneas an epic figure with a taste for voyaging, and one who, as Tennyson imagined him at the end of his life, planned a final spectacular journey. Let Ulysses have the last word:

> There lies the port; the vessel puffs her sail:
> There gloom the dark, broad seas. My mariners,
> Souls that have toiled, and wrought, and thought with me—
> That ever with a frolic welcome took
> The thunder and the sunshine, and opposed
> Free hearts, free foreheads—you and I are old;
> Old age hath yet his honor and his toil;
> Death closes all; but something ere the end,
> Some work of noble note, may yet be done,
> Not unbecoming men that strove with Gods.
> The lights begin to twinkle from the rocks:
> The long day wanes: the slow moon climbs: the deep
> Moans round with many voices. Come, my friends.
> 'Tis not too late to seek a newer world.

AFTERWORD

Some of the spaceflight pioneers discussed in this book are now working behind the scenes to lay out the foundations for a non-profit organization to pursue the challenge of interstellar flight in earnest. The Interstellar Flight Foundation is a nonprofit (501c3) corporation primarily supported through philanthropic donations that conducts visionary, credible research toward the grand challenge of interstellar flight. Appreciative of the fact that its goal is beyond currently foreseeable solutions, the Foundation aims to demonstrate incremental progress over the course of this long journey. In addition to the research itself, the foundation will also address public education and will include provisions for capitalizing on any marketable spin-offs that are discovered along the way.

If you are interested in contacting the Foundation, please do so through me at gilster@centauri-dreams.org.

For updates on the issues discussed in this book, as well as news on technical developments in deep-space exploration and ongoing research on interstellar exploration, please visit my website, www.centauri-dreams.org.

NOTES

CHAPTER 1

p. 1: "The only thing left was rockets!"—Interview with Ray Lewis at Marshall Space Flight Center, July 30, 2003.

p. 2: "all written in the dry, precise cadences of physics."—Like so many of these papers, however, the terse prose belies the staggering possibilities that grow from these ideas. See, for example, Lewis's collaborative paper "Antiproton-Catalyzed Microfission/Fusion Propulsion Systems for Exploration of the Outer Solar System and Beyond," written with G. Gaidos, Gerald Smith, B. Dundore, and S. Chakrabarti, in *Space Technology and Applications International Forum-1997*, edited by Mohamed S. El-Genk (New York: American Institute of Physics Press, 1997), p.1499; also available online at http://www.engr.psu.edu/antimatter/Papers/ICAN.pdf. A later chapter discusses Lewis and his colleagues' concept of AIMStar, a propulsion system for interstellar precursor missions.

p. 2: "mirror matter."—Indeed, the best general introduction to antimatter is a book by Robert Forward and Joel Davis of the same name, *Mirror Matter: Pioneering Antimatter Physics* (New York: John Wiley & Sons, 1988). Forward was also editor for several years of an influential antimatter newsletter that connected theorists and experimenters in a field that continues to flourish.

p. 4: "we have no alternatives."—Interview with John Cole at Marshall Space Flight Center, July 30, 2003.

p. 5: "to help determine latitude."—Ben R. Finney, "The Prince and the Eunuch," in Ben Finney and Eric M. Jones, eds. *Interstellar Migration and the Human Experience* (Berkeley: University of California Press, 1985), a collection of papers from a 1983 conference discussed in Chapter 10.

p. 7: "a full cargo of peas."—As found in Richard Hinckley Allen, *Star Names: Their Lore and Meaning* (New York: Dover Publications, 1963), a reprint of the charming 1899 original *Star-Names and Their Meanings*.

p. 8: "260,000 AU."—This is NASA information, drawn from a Web site about solar system distances, http://heasarc.gsfc.nasa.gov/docs/cosmic/solar_system_info.html.

p. 8: "(not to mention reaping plenty of airline miles)."—Aczel's useful comparison appeared as "A Measurement Whose Time Has Come" in the *New York Times*, September 9, 2003, p. F3.

p. 8: "seven miles apart."—Terrile's analogy comes from an interview Geoffrey Landis conducted with three scientists who are also science fiction writers, David Brin, Robert Forward, and Jonathan Vos Post, in the magazine *Science Fiction Age*. "Starflight Without Warp Drive" is now available on the Web at http://www.sff.net/people/Geoffrey.Landis/stl.htp. I have used 100 billion as the total number of stars in the Galaxy, although some estimates range as high as 500 billion.

p. 9: "at least one miracle in development in order to enable an interstellar mission."—Howe is the author of the NASA Institute for Advanced Concepts study "Antimatter Driven Sail for Deep Space Exploration." Howe's work can be found on the NIAC Web site at www.niac.usra.edu.

p. 10: "Pluto....an average 40 astronomical units (AU) from the Sun."—Pluto's 248-year orbit is highly elliptical, and varies from 29.6 AU to 48.9 AU. For about twenty years per orbit, Pluto actually passes within the orbit of Neptune, the only planet to cross the orbit of another.

p. 10: "fully three-quarters the size of Pluto."—The team that discovered Sedna at California's Palomar Observatory speculates that the planetoid may actually belong to the cloud of cometary debris known as the Oort Cloud. See "Planetoid on the Fringe," in *New Scientist* 165 (March 20, 2004), p. 179.

p. 10: "surprisingly thick band." This prediction is found in Davies's book *Beyond Pluto: Exploring the Outer Limits of the Solar System* (Cambridge, U.K.: Cambridge University Press, 2001), p. 149.

p. 12: "in semiconductors, materials, simulation, propulsion."—As reported in "Goldin Announces Interstellar Probe Within 25 Years," Pathfinder Special Reports, Aerospace FYI, July 3, 1997, a document found among Robert Forward's papers.

p. 12: "developed at the Jet Propulsion Laboratory in 1977."—L. D. Jaffe et al., "An Interstellar Precursor Mission," JPL Publication 77-70, Jet Propulsion Laboratory, Pasadena, California, October 30, 1977.

p. 12: "out to perhaps 1,000 AU."—The Interstellar Precursor Mission is summarized in L. D. Jaffe et al., "An Interstellar Precursor Mission," *Journal of the British Interplanetary Society* 33 (1980): 3–26.

p. 12: "to transmit its findings,"—See J. R. Lesh, C. J. Ruggier, and R. J. Cesarone, "Space Communications Technologies for Interstellar Missions," *Journal of the British Interplanetary Society* 49 (1996), pages 7–8 for a recap of the Interstellar Precursor and TAU missions. On the TAU mission itself, see K. T. Nock, "TAU—A Mission to a Thousand Astronomical Units," 19th AIAA/DGLR/JSASS International Electric Propulsion Conference paper AIAA-87-1049, Colorado Springs, Colo., May 11–13, 1987.

p. 12: "also remained an option."—See Gregory Matloff, *Deep Space Probes* (Chichester, U.K.: Praxis Publishing, 2000), pages 26–27 for more details on the TAU mission.

p. 13: "if we effectively utilize hybridized technologies."—Daniel S. Goldin, "Remarks as Prepared for Presentation to the 100th Anniversary Meeting of the American Astronomical Society," June 3, 1999. Available online at https://www.aas.org/policy/1999/GoldinTalkChicago.html.

p. 14: "have appeared in the Congressional Record."—From Forward's "Fast Forward Fifty Years," an unfinished manuscript available only on the Web at http://www.robertforward.com/Fast_Forward_Fifty_Years.htm. This was to have been Forward's autobiography, but he did not live to complete it.

p. 14: "10 to 20 years later."—Robert L. Forward, "A National Space Program for Interstellar Exploration," *Future Space Programs* 1975, vol. VI, Subcommittee on Space Science and Applications, Committee on Science and Technology, U.S. House of Representatives, Serial M, 94th Congress (September, 1975). Forward's roadmap for interstellar exploration is evaluated and placed in context by Saul J. Adelman and Benjamin Adelman in their book *Bound for the Stars* (Englewood Cliffs, N.J.: Prentice-Hall, 1981), pages 290–308.

p. 15: "pass on that information to the public."—Forward, "Fast Forward."

p. 16: "the gravity focus."—Gregory Matloff, "Solar Sailing for Radio Astronomy and SETI: An Extrasolar Mission to 550 AU," *Journal of the British Interplanetary Society* 47 (1994): 476–84.

p. 17: "*Flight to the Stars*."—Strong's book remains an essential one, despite its dated examinations of many propulsion concepts. *Flight to the Stars* was published by London's Temple Press in 1965 and remains a hard to find, much sought after tome in used book shops.

p. 17: "excess fusion energy."—Gregory L. Matloff and H. H. Chiu, "Some Aspects of Thermonuclear Propulsion," *Journal of Astronautical Sciences* 18 (July/August 1970): 57–62.

p. 17: "*Orphans of the Sky*"—"Universe" appeared in the May, 1941, issue of *Astounding Science Fiction*. *Orphans of the Sky* folded it and another story, "Common Sense," into a novel in 1963.

p. 18: "much closer to home."—Interview with Gregory Matloff, August 18, 2003.

p. 19: "but it can no longer be considered impossible."—Forward made this comment on numerous occasions. The paper I quote from is his "Ad Astra!" in the *Journal of the British Interplanetary Society* 49 (1996), page 23.

p. 20: "the *Journal of Spacecraft and Rockets* in 1984,"—Robert L. Forward, "Roundtrip Interstellar Travel Using Laser-Pushed Lightsails," *Journal of Spacecraft* 21, no. 2 (March/April 1984): 187–95.

p. 23: "hundreds of years to get there."—Interview with Humphrey Price at the Jet Propulsion Laboratory, January 31, 2003.

p. 24: "in less than 50 years."—Ralph McNutt, "A Realistic Interstellar Explorer," NASA Institute for Advanced Concepts Phase 1 Final Report, 1999.

p. 24: "for the last thousand."—This observation, and many other thoughts on artifacts and survival, can be found in Stewart Brand's wonderful *The Clock of the Long Now* (New York: Basic Books, 1999). To make his case for long-term thinking, Brand adds a zero to our conventional dating, a reminder that we'll need to reset computers in the year 10,000 just as we did in the era of Y2K. 2004 thus becomes, in Brand's notation, 02004.

p. 24: "within our reach."—McNutt, "Interstellar Explorer."

p. 24: "Earth-like planet."—See especially Stephen H. Dole, *Habitable Planets for Man* (New York: Blaisdell Publishing/Ginn and Company, 1964), a study prepared for the RAND Corporation. More widely available is Dole and Isaac Asimov's *Planets for Man* (New York: Random House, 1964), which is based upon the more technical title.

p. 25: "it made as much difference as it did."—Telephone interview with Ralph McNutt, January 23, 2003.

p. 25: "great instruments of civilization."—Brand, *The Clock*, 103.

p. 26: "an incredibly ballsy thing to do!"—Interview with Geoffrey Landis at the Glenn Research Center, April 3, 2003.

p. 27: "*Tales from the White Hart.*"—Arthur C. Clarke, *Tales from the White Hart* (New York: Ballantine Books, 1957).

p. 28: "for a very long time now."—Landis interview, April 3, 2003.

p. 29: "landmarks of interstellar research,"—A. R. Martin, ed., *Project Daedalus Final Report.* Supplement to the *Journal of the British Interplanetary Society*, 1978.

CHAPTER 2

p. 32: "fourth millennium B.C. temples there."—As noted in Richard Hinckley Allen, *Star Names: Their Lore and Meaning* (New York: Dover Publications, 1963), p. 153 (reprint of the 1899 original). Also noted in "Sounds of a Star: Acoustic Oscillations in Solar-Twin 'Alpha Cen A' Observed from La Silla by Swiss Team," press release from the European Southern Observatory, June 28, 2001.

p. 34: "cooler than the Sun."—For a quick view of stellar types, see Robert Zubrin's *Entering Space: Creating a Spacefaring Civilization* (New York: Tarcher/Putnam, 1999), pages 231–33. For a more detailed study, see William J. Kaufman's *Universe* (New York: W. H. Freeman, 1985), a textbook with much background information.

p. 34: "seven times smaller than the Sun."—As reported by Robert Roy Britt in "Detailed Measures Taken of Closest Star System, Alpha Centauri," posted on www.space.com on March 17, 2003.

p. 35: "ranging from –18.1 to –20.6."—Drawn from information developed by Robert J. Sawyer, who based it on the work of Edward F. Guinan of the Department of Astronomy and Astrophysics at Villanova University. Sawyer used these orbital mechanics in his science fiction novel *Illegal Alien* (New York: Ace Books, 1997).

p. 36: "an independent star happening to pass close by."—Asimov discusses the visual effects of the Centauri system in his *Alpha Centauri* (New York: Lothrop, Lee & Shepard Co., 1976), 101–105.

p. 36: "within the Alpha Centauri system."—For more on Proxima's radial velocity and its implications, see Robert Matthews and Gerard Gilmore, "Is Proxima Really in Orbit About Alpha CEN A/B?" *Royal Astronomical Society Monthly Notices* 261, no. 2 (1993), pages L5-L7.

p. 36: "cannot be ruled out at least around Alpha Centauri A."—See M. Barbieri, F. Marzari, and H. Scholl, "Formation of Terrestrial Planets in Close Binary Systems: the Case of Alpha Centauri A," *Astronomy & Astrophysics* 396 (2002): 219–24. Available online at http://arxiv.org/abs/astro-ph/0209118.

p. 36: "stable planetary orbits within binary star systems."—Alan Hale, "Nearby Solar-Type Stars as Candidates for Interstellar Robotic Missions," *Journal of the British Interplanetary Society* 49 (1996), 150–54.

p. 37: "Paul Wiegert and Matt Holman in 1997."—Paul Wiegert and Matt Holman, "The Stability of Planets in the Alpha Centauri System," *Astronomical Journal* 113 (1997): 1445–50.

p. 37: "and, possibly, could harbor life."—Barbieri, Marzari, and Scholl, "Formation of Terrestrial Planets."

p. 38: "data on the star field around Proxima."—G. F. Benedict et al., "Searching for Planets Near Proxima Centauri: A Status Report." *Bulletin of the American Astronomical Society* 26 (1994): 930.

p. 38: "roughly half the distance between Earth and the Sun."—A. B. Schultz et al., "A Possible Companion to Proxima Centauri," *Astronomical Journal* 115 (1998): 345–50.

p. 38: "about 80 percent of the mass of Jupiter in nearby orbits."—G. F. Benedict et al., "Interferometric Astrometry of Proxima Centauri and Barnard's Star Using Hubble Space Telescope Fine Guidance Sensor 3: Detection Limits for Substellar Companions," *Astronomical Journal* 118 (1999): 1086–1100.

p. 38: "a fraction of the distance between Mercury and the Sun."—See Michael Endl et al., "Extrasolar Terrestrial Planets: Can We Detect Them Already?" *Scientific Frontiers in Research on Extrasolar Planets, ASP Conference Series,* vol. 294 (2003). The answer, incidentally, is ". . . not yet, but we get pretty close. . ." For more on habitable zones, see J. F. Kasting, D. P. Whitmire, and R. T. Reynolds, "Habitable Zones Around Main Sequence Stars," *Icarus* 101 (1993): 108–128.

p. 39: "warm enough to prevent its gases from freezing out."—The Ames work is described in Manoj Joshi, Robert Haberle, and R. Reynolds, "Simulations of the Atmospheres of Synchronously Rotating Terrestrial Planets Orbiting M Dwarfs: Conditions for Atmospheric Collapse and the Implications for Habitability," *Icarus* 129 (1997), pages 450–65. See also Martin J. Heath et al., "Habitability of Planets Around Red Dwarf Stars," *Origins of Life and Evolution of the Biosphere* 29 (1999), pages 405–24. For a fine overview of these findings and a discussion of related work, see Ken Croswell's "Red, Willing and Able," in *New Scientist* 169 (January, 2001), pages 28–31.

p. 39: "a definitive conclusion."—As reported in Robert Roy Britt, "Report of Earth-Sized Planet Around Another Star Premature." Posted on www.space.com January 19, 2001. For details, see L. R. Doyle et al., "From CM Draconis to the Crowded Field BW3: Aspects of the Search for Extrasolar Planets Around Small Eclipsing Binaries," *Proceedings of Bioastronomy 99: A new Era in Bioastronomy,* eds. G. M. Lemarchand and K. J. Meech, ASP Conf. Ser. vol. 213, (2000) p. 159.

p. 40: "together with the whole system of life forms on which they depend."—Stephen H. Dole and Isaac Asimov, *Planets for Man* (New York: Random House, 1964), 108. Dole's RAND Corporation study is titled *Habitable Planets for Man* (New York: Blaisdell, 1964).

p. 40: "for any star or star system on our list . . ."—Ibid., 178.

p. 41: "and not so far toward the edge to be metal-poor."—The issue was raised by Peter D. Ward and Donald Brownlee in their book *Rare Earth* (New York: Copernicus Books, 2000).

p. 41: "so-called 'hot Jupiters' orbiting their stars in orbits tighter than this."—Matloff tells this story in his *Deep Space Probes* (Chichester, U.K.: Praxis Publishing, 2000), 21.

p. 41: "28 percent the distance of Mercury from the Sun."—The 55 Cancri planet was announced in Paul Butler et al., "Three New 51 Pegasi-Type Planets," *Astrophysical Journal* 474 (1997): L115–L118.

p. 42: "little effect on the possible planetary system of the other." — Press release, "McDonald Observatory Planet Search Finds First Planet Orbiting Close-in Binary Star," University of Texas, October 9, 2002.

p. 43: "It was a fantastic moment." — An interview with Geoff Marcy conducted by the Jet Propulsion Laboratory's media team and available on its Planet Quest Web site at http://planetquest.jpl.nasa.gov/news/marcy.html.

p. 43: "exotic far-away places like Brooklyn." — Daniel S. Goldin, "Remarks as Prepared for Presentation to the 100th Anniversary Meeting of the American Astronomical Society," June 3, 1999. Available online at www.aas.org/policy/1999/GoldinTalkChicago.html.

p. 44: "circling the star HD70642 about every six years." — Press release, "Astronomers Find 'Home from Home' — 90 Light Years Away," Particle Physics and Astronomy Research Council, July 3, 2003. See also Tariq Malik, "Celestial Soulmate? Jupiter-Like Planet Found in System Similar to Ours," www.space.com, July 3, 2003.

p. 44: "later extended by them with the discovery of a second planet." — Press release, "UC Berkeley Astronomers Find Jupiter-Sized Planet Around Nearby Star in Big Dipper," University of California at Berkeley, August 15, 2001.

p. 45: "it should be possible to obtain its image." — W. B. Sparks et al., "Detection of Planets with the Hubble Space Telescope Advanced Camera for Surveys," from A New Era in Bioastronomy, 6th Bioastronomy Meeting (August 2–6, 1999). Available online at http://acs.pha.jhu.edu/instrument/papers/documents/acs_planets.pdf.

p. 45: "in the habitable zones around other solar systems." — Ralph McNutt, of the Johns Hopkins University Applied Physics Laboratory, provided an overview of Terrestrial Planet Finder in his presentation "Space Exploration Beyond 2020," given at the Military and Aerospace Programmable Logic Device International Conference, 1999, and available on the Web at http://klabs.org/richcontent/MAPLDCon99/Presentations/A0_McNutt_S.pdf.

p. 46: "by examining their spectral signatures." — For more on Life Finder, see University of Arizona astronomer Neville Woolf's study for NASA's Institute for Advanced Concepts, "Life Finder: Very Large Optics for the Study of Extra-Solar Terrestrial Planets." Available at the NIAC Web site www.niac.usra.edu.

p. 46: "as a colossal interferometer that combines the image from each." — From press release, "Powerful X-Ray Astronomy Telescope Should Lead to Black Hole Exploration," University of Colorado at Boulder (September 13, 2000).

p. 46: "the size of an automobile at the center of the Milky Way." — For a detailed look at Cash's x-ray interferometer, see his "X-Ray Interferometry: Ultimate Astronomical Imaging," the final report of a Phase II study for NASA's Institute for Advanced Concepts (April, 2002). Both Phase I and II reports are available online at www.niac.usra.edu.

p. 47: "some scientists at Princeton who know how to do that." — This and the following quotes from Webster Cash are from a telephone interview with the astronomer on August 29, 2003.

p. 49: "working with 2-meter telescopes to examine infrared light." — Betsy Mason, "Look of Life," *New Scientist* 179 (July 12, 2003): 28.

p. 50: "within a range of 21 light years." — These figures are drawn from Gregory Matloff and Eugene Mallove in their essential *Starflight Handbook*

(New York: John Wiley & Sons, 1989), and adjusted for the discoveries in 1997, 2002, and 2003 of three more stars within the 21-light year range. The newly discovered stars were all M-class dwarfs save for a single white dwarf. Matloff and Mallove go on to explain their choice of a 21-light year zone: "A starship encountering a solar system 21 ly removed on the day of birth of an earth-bound astronomer would radio back initial scientific data that will be newly received just in time for the starchild to analyze for her senior thesis in college!" (p. 27).

p. 50: "within a sphere with a radius of 70 light years."—G. Vulpetti, "Problems and Perspectives in Interstellar Exploration," *Journal of the British Interplanetary Society* 52 (1999): 314.

p. 50: "to create acid rain, ozone loss, and global winter."—This theory was first suggested by astrophysicist Richard Muller and colleagues in an article in *Nature*. See M. Davis, P. Hut, and R. A. Muller, "Extinctions of Species by Periodic Comet Showers," *Nature* 308 (1984): 715–17. The same issue saw an independent article proposing much the same theory: D. P. Whitmire and A. A. Jackson, "Are Periodic Mass Extinctions Driven by a Distant Solar Companion?" (pp. 713–15). Research on Nemesis has been summarized by Donald Goldsmith in his *Nemesis: The Death-Star and Other Theories of Mass Extinction* (New York: Walker and Company, 1985). See also Richard Muller's *Nemesis: The Death Star* (New York: Weidenfeld and Nicolson, 1988).

p. 50: "by Edward Emerson Barnard, working at California's Lick Observatory."—E. E. Barnard, "A Small Star with Large Proper Motion," *Astronomical Journal* 29 (1916): 181–83.

p. 51: "and Groombridge 1618 (a K star 15 light years away)."—John H. Mauldin offers a good discussion of local targets in his *Prospects for Interstellar Travel*, from which these numbers are drawn. The book is Volume 80 in the American Astronautical Society's Science and Technology Series (San Diego, Calif.: Univelt, 1992). The discussion on nearby stars begins on p. 139.

p. 52: "and grains of sand begin to look like torpedoes."—Charles Pellegrino, *Flying to Valhalla* (New York: William Morrow and Co., 1993).

p. 52: "some scientists argue that Voyager 1 has already reached the termination shock, a view that is still controversial."—Two papers published in late 2003 define the debate: see S. M. Krimigis et al., "Voyager 1 Exited the Solar Wind at a Distance of ~85 AU from the Sun," *Nature* 426 (2003): 45–48; and F. B. McDonald et al., "Enhancements of Energetic Particles Near the Heliospheric Termination Shock," *Nature* 426 (2003): 48–51. The two papers come to different conclusions using the same data, suggesting that the shape of the termination shock may be more complicated than was originally supposed.

p. 53: "measuring its impact upon the spacecraft's skin with its plasma wave instrument."—For more on interplanetary dust, see Eberhard Grun, Harald Kruger, and Markus Landgraf, "Dust Measurements in the Outer Solar System," http://arXiv.org, astro-ph/9902036 v 1 (February 2, 1999). For the specifics on Voyager's dust measurements, see D. A. Gurnett et al., "Micron-sized Dust Particles Detected in the Outer Solar System by the Voyager 1 and 2 Plasma Wave Instruments," *Geophys. Res. Lett.* 24 (1997): 3125–28.

p. 54: "any entering ship will be blasted."—Telephone interview with Stephen D. Howe, April 22, 2003.

p. 54: "—our solar system seems to be in a pocket of unusually sparse material)."—A. R. Martin writes about interstellar dust concentrations in "Bombardment by Interstellar Material and Its Effects on the Vehicle," *Project Daedalus Final Report (Journal of the British Interplanetary Society*, 1978): S116–S121.

p. 55: "might have made the formation of life impossible."—These speculations are from Mauldin, *Prospects*, 102.

p. 55: "to protect the vehicle."—Martin, "Bombardment," S119.

p. 56: "during the few hours available for planetary observation."—Alan Bond, "Project Daedalus: Target System Encounter Protection," in *Project Daedalus Final Report (Journal of the British Interplanetary Society*, 1978), S123–S125.

p. 56: "to deflect or destroy the object."—Matloff and Mallove, *Starflight*, 170.

CHAPTER 3

p. 60: "carrying 100 pounds of fuel."—An excellent discussion of the rocket problem can be found in Bernard Haisch and Alfonso Rueda, "Prospects for an Interstellar Mission: Hard Technology but Surprising Physics Possibilities," *Mercury* 29, Issue 4 (July/August 2000), p. 26. The article draws on ideas presented at a July, 1998, workshop at Caltech called "Robotic Interstellar Exploration in the Next Century." The conference was sponsored by the Advanced Concepts Office of the Jet Propulsion Laboratory and the Office of Space Science at NASA headquarters.

p. 60: "three hundred million supertankers to make the 100-year journey and stop!"—From an interview with Marc Millis at Glenn Research Center in Cleveland, April 3, 2003. Millis also discussed these numbers in his article "Breaking Through to the Stars," which ran in *Ad Astra: The Magazine of the National Space Society* 9, no. 1 (Jan./Feb. 1997), pp. 36–40. A revised version of this article can be found online at http://www.grc.nasa.gov/WWW/bpp/bpp_INTERSTELLAR.htm.

p. 60: "the equations that relate mass to velocity change."—Robert Zubrin provides an analysis of the "rocket equation" in *Entering Space: Creating a Spacefaring Civilization* (New York: Tarcher/Putnam, 1999), p. 36. See also Eugene Mallove and Gregory Matloff, *The Starflight Handbook: A Pioneer's Guide to Interstellar Travel* (New York: John Wiley & Sons, 1989), pp. 38–40.

p. 61: "for its asteroid encounter."—From Marc Rayman's mission log on Deep Space 1 (August 22, 1999 entry), posted online at http://nmp.jpl.nasa.gov/ds1/arch/mrlogr.html. Rayman was chief mission engineer and deputy mission manager for Deep Space 1.

p. 62: "channeled by magnetic fields to produce thrust."—Duncan Graham-Rowe, "Nuclear Fusion Could Power NASA Spacecraft," *New Scientist* 23, January 2003.

p. 63: "chemical propulsion is pretty non-optimal.""—Telephone interview with Les Johnson, January 6, 2003.

p. 63: "a four-month return voyage."—Steven D. Howe, "Reducing the Risk to Mars: The Gas Core Nuclear Rocket," Space Technology and Applications International Forum, Jan. 1998. Available online at http://internet.cybermesa.com/~mrpbar/staifpaper.html. Howe also explores nuclear propulsion in "Nuclear Rocket to Mars," *Aerospace America*, Aug. 2000, p. 39. See also his "High Energy-Density

Propulsion: Reducing the Risk to Humans in Planetary Exploration," *Space Policy* 17, Issue 4 (November 2001): 275–83.

p. 64: "a prime target for future astronomy."—MITEE is discussed on Plus Ultra's Web site at www.newworlds.com/mitee.html.

p. 65: "radiating waste reactor heat into space."—Leonard David, "Prometheus: The Paradigm Buster," www.space.com, July 2, 2003.

p. 65: "this is stadium lighting, kilowatts instead of watts."—As quoted in "Nukes in Space," *U.S. News & World Report,* April 28, 2003, p. 55.

p. 66: "a dual shock-absorber system."—George Dyson tells the Orion story—one in which his father, Freeman Dyson, played a huge part—in *Project Orion: The True Story of the Atomic Spaceship* (New York: Henry Holt & Co., 2002).

p. 67: "rich in ice and hydrocarbons."—Ibid., p. 191.

p. 67: "to navigate through the shoals of three federal agencies."—Freeman Dyson provides this and much more information about Orion in *Disturbing the Universe* (New York: Harper and Row, 1979).

p. 68: "the Strategic Defense Initiative in the 1980s."—Freeman Dyson wrote the obituary for Orion in a classic article called "Death of a Project: Research Is Stopped on a System of Space Propulsion Which Broke All the Rules of the Political Game," *Science* 149, no. 3680 (July 9, 1965), p. 141.

p. 68: "with arrival at Alpha Centauri in 130 years."—Freeman Dyson, "Interstellar Transport," *Physics Today*, Oct. 1968.

p. 68: "without inventing entirely new technologies."—As noted by Geoffrey Landis in "The Ultimate Exploration: A Review of Propulsion Concepts for Interstellar Flight," presented at the American Association for the Advancement of Science annual meeting, February 2002.

p. 69: "as part of its defense against alien invaders."—Larry Niven and Jerry Pournelle, *Footfall* (New York: Ballantine Books, 1987).

p. 69: "Both Poul Anderson and Stephen Baxter have worked entertainingly with the idea."—Anderson in *Orion Shall Rise* (New York: Pocket Books, 1983), Baxter in *Manifold: Space* (New York: Del Rey, 2001).

p. 69: "after which it is safe to explode a series of atomic devices next to it for propulsion."—Vernor Vinge, *Marooned in Realtime* (New York: Baen Books, 1989).

p. 69: "*2001: A Space Odyssey.*"—The comment on Kubrick and quote from Taylor are from "Deep Impact: Filming a Cosmic Catastrophe," *The Planetary Report*, 18, no. 3 (May/June 1998), pp. 12–15.

p. 69: "against the ship's thick ablative tail plate."—As quoted in George Dyson, *Project Orion*, 271.

p. 69: "that would be dissolved in ordinary water."—Robert M. Zubrin, "Nuclear Salt Water Rockets: High Thrust at 10,000 sec ISP." *Journal of the British Interplanetary Society* 44 (1991), 371–76.

p. 70: "a close-up look at our neighboring star systems."—John Cramer provides a summary of Zubrin's ideas in one of his science columns for *Analog Science Fiction and Fact,* "Nuke Your Way to the Stars," in the mid-December, 1992, issue of the magazine. This column can also be found online at http://www.npl.washington.edu/AV/altvw56.html.

p. 71: "whereas the neutrons produced by the deuterium/tritium reaction cannot."—Landis, "The Ultimate Exploration," 4.

p. 71: "where collisions between them are likely and fusion possible."—See Zubrin, *Entering Space,* pp. 84–85 for a good explanation of fusion concepts.

p. 71: "to keep the hot plasma from contacting the chamber's walls."—Dwain Spencer, "Fusion Propulsion for Interstellar Missions," *Annals of the New York Academy of Science* 140 (December 1966): 407–18. Spencer's paper is one of the first to discuss practical fusion concepts in relation to interstellar flight.

p. 72: "the engine would use lasers to ignite fusion in the pellets."—R. Hyde, L. Wood, and J. Nuckolls, "Prospects for Rocket Propulsion with Laser-Induced Fusion Micropellets," *American Institute of Aeronautics and Astronautics Paper* 72-1063 (November 1972). The Hyde design is also discussed in Saul J. Adelman and Benjamin Adelman, *Bound for the Stars* (Englewood Cliffs, N.J.: Prentice-Hall, 1981), pages 195–96.

p. 72: "who foresees the use of helium-3 as a fuel source for the twenty-first century."—As quoted in Julie Wakefield, "Moon's Helium-3 Could Power Earth," www.space.com, June 30, 2000.

p. 72: "a key paper by Friedwardt Winterberg"—F. Winterberg, "Rocket Propulsion by Thermonuclear Micro-Bombs Ignited with Intense Relativistic Electron Beams," *Raumfahrtforschung* 15 (1971): 208–17.

p. 72: "initiating fusion in small amounts of high density materials with lasers,"—J. Nuckolls et al., "Laser Compression of Matter to Super-High Densities: Thermonuclear Applications," *Nature* 239 (1972): 139–42.

p. 73: "along with 20,000 tons of deuterium."—Within the British Interplanetary Society final report on Project Daedalus are the two primary articles on its propulsion system. These are A. R. Martin and A. Bond, "Project Daedalus: The Propulsion System—Part I: Theoretical Considerations and Calculations," pp. S44–S62; and the same authors' "Project Daedalus: The Propulsion System—Part II: Engineering Design Considerations and Calculations," pp. S63–S82. Martin and Bond review ICF concepts in a later paper, "Nuclear Pulse Propulsion: A Historical Review of an Advanced Propulsion Concept," *Journal of the British Interplanetary Society* 32 (1979), pp. 283–310.

p. 73: "rather than a heroic effort on the part of a planet-bound civilization."—R. C. Parkinson, "Project Daedalus: Propellant Acquisition Techniques," in *Project Daedalus Final Report* (1978), pp. S83–S89.

p. 73: "dense enough—and hot enough—to light the fusion reaction."—A good overview of inertial confinement fusion concepts is found in Brice N. Cassenti's "Nuclear Pulse Propulsion for Interplanetary Travel," *American Institute of Physics Conference Proceedings*, vol. 552 (1), (February 2, 2001): 881–85.

p. 73: "Very tough."—Telephone interview with Terry Kammash, August 14, 2003.

p. 74: "their inner walls coated with fusion fuel."—See T. Kammash and D. L. Galbraith, "A High Gain Fusion Reactor Based on the Magnetically Insulated Inertial Confinement Fusion (MICF) Reactor," *Nuclear Fusion* 29 (1989), pp. 1079–1099.

p. 74: "replacing the laser with the most exotic material known on Earth: antimatter."—Kammash developed this concept in a study for NASA's Institute for Advanced Concepts called "Antiproton Driven Magnetically Insulated Inertial Confinement Fusion (MICF) Propulsion System," available at the NIAC Web site, www.niac.usra

CHAPTER 4

p. 77: "the few ragged survivors of the 'antimatter wars' of 16 billion years ago."—John G. Cramer, "Antimatter in a Trap," *Analog Science Fiction and Fact*, December 1985.

p. 77: "at the blinding moment of collision."—Paul Preuss has written one of the few science fiction novels that deals with the production of antimatter in realistic settings. The novel, called *Broken Symmetries* (Hastings-on-Hudson, N.Y.: Ultramarine Publishing, 1983), deals with particle research and the odd things that come out of collisions at relativistic speeds.

p. 78: "delivered in a lecture at the Fourth International Astronautical Congress and subsequently published,"—E. Sänger, "The Theory of Photon Rockets," in *Space Flight Problems* (Biel-Bienne, Switzerland: Laubscher, 1953).

p. 78: "or use the flow to heat a propellant such as liquid hydrogen."—Robert L. Forward and Joel Davis, *Mirror Matter: Pioneering Antimatter Physics* (New York: John Wiley & Sons, 1988), 135–36. For the Morgan design, see p. 239.

p. 79: "770 pounds of antimatter and 24 tons of liquid hydrogen."—Ibid., 239–40.

p. 79: "one hundred times more efficient than a fusion reaction."—Eugene Mallove and Gregory Matloff, *The Starflight Handbook: A Pioneer's Guide to Interstellar Travel* (New York: John Wiley & Sons, 1989), 51.

p. 80: "although the antimatter theme was by then but a minor one."—Forward's *Mirror Matter* has a nice overview of science fictional treatments of antimatter on pages 49–61.

p. 81: "the related problem of building a workable, long-life nuclear-fusion engine."—The AIMStar mission is described in Kevin J. Kramer et al., "AIMStar: Antimatter Initiated Microfusion for Pre-Cursor Interstellar Missions," in *Space Technology and Applications International Forum-2000*, edited by M. S. El-Genk, CP504, 1412–19.

p. 82: "the pellets could be exploded at the rate of one every second to obtain the needed thrust."—G. Gaidos, R. A. Lewis, and G. A. Smith, "Antiproton Catalyzed Microfission/Fusion Propulsion Systems for Exploration of the Outer Solar System and Beyond," available at the Penn State Web site:
http://www.engr.psu.edu/antimatter/Papers/ICAN.pdf. A popularized version of the Penn State work can be found in Stefano Coledan, "Antimatter Spaceships," *Popular Mechanics*, February 2003.

p. 83: "studies that would increase these capabilities by 100 times."—"Reaching for the Stars: Scientists Examine Using Antimatter and Fusion to Propel Spacecraft," Science@NASA, April 12, 1999. Available online at
http://science.nasa.gov/newhome/headlines/prop12apr99_1.htm.

p. 83: "can store antiprotons for up to ten days."—M. H. Holzscheiter et al., "Production and Trapping of Antimatter for Space Propulsion Applications," available online at
http://www.engr.psu.edu/antimatter/papers/anti_prod.pdf. Also see Penn State's Web site at
http://www.engr.psu.edu/antimatter/introduction2.html.

p. 83: "studies that would increase these capabilities by 100 times."—G. Jackson, "Commercial Production and Use of Anti-Protons," *Proceedings* *of EPAC 2002*, Paris, France. Available online at http://accelconf.web.cern.ch/AccelConf/e02/PAPERS/FRXGB003.pdf.

p. 83: "up to 300 times the thrust of a conventional chemical rocket engine."—Emrich's work is described in "Nuclear Fusion Could Power NASA Spacecraft," by Duncan Graham-Rowe in *New Scientist*, January 23, 2003.

p. 83: "that avoids the instabilities of older containment systems."—See Terry Kammash and M. J. Lee, "A Fusion Propulsion System for Near-Term Space Exploration," *Journal of the British Interplanetary Society* 49 (1996), pp. 351–56 for an explanation of the gas dynamic mirror concept.

p. 84: "while Jupiter would be only a month away."—David Dooling, "NASA to Begin Fusion Reactor Testing," space.com, July 21, 2000.

p. 85: "*Honor Bound Honor Born*"—Steven D. Howe, *Honor Bound Honor Born* (Los Alamos, N.M.: LunaTech Press, 1997).

p. 85: "Antimatter Driven Sail for Deep Space Missions."—Steven D. Howe, "Antimatter Driven Sail for Deep Space Missions," a Phase I study for NASA's Institute for Advanced Concepts, available online at http://www.niac.usra.edu/studies.

p. 86: "flesh out the entire architecture for both missions, the Oort Cloud and Alpha Centauri."—Telephone interview with Stephen D. Howe, April 22, 2003.

p. 87: "make a 100-watt bulb shine for fifteen minutes."—From CERN's "Frequently Asked Questions About Antimatter," available online at http://athena-positrons.web.cern.ch/ATHENA-positrons/wwwathena/FAQ.html.

p. 87: "prevent it from contacting the chamber walls and annihilating."—Steven D. Howe and Gerald A. Smith, "Enabling Exploration of Deep Space: High Density Storage of Antimatter," NIAC Phase 1 Final Report. Available at the NIAC Web site: www.niac.usra.edu.

p. 88: "so that it drifts toward the sail."—Howe interview, April 22, 2003.

p. 88: "he is still in the realm of where we can have human exploration of the outer planets."—Interview with John Cole at Marshall Space Flight Center, July 30, 2003.

p. 88: "announced that they had created, for the first time, large numbers of antihydrogen atoms."—From CERN press release, "Thousands of Cold Anti-Atoms Produced at CERN," September 18, 2002. The CERN experiments with a great deal of supporting information are presented online at CERN's Web site: http://athena.web.cern.ch/athena.

p. 89: "ten million dollars per milligram."—Robert L. Forward, *Indistinguishable from Magic* (New York: Baen Books, 1995), 25–26.

p. 91: "robotic probes and even manned missions to neighboring star systems."—Advanced antimatter concepts are from "Advanced Propulsion Concepts," a CD compiled in 1989 by the Jet Propulsion Laboratory and made available online at http://www.islandone.org/APC.

p. 91: "such a rocket pushes engineering to the breaking point."—The story, like most of Smith's, is a gem. "Golden the Ship Was—Oh! Oh! Oh!" appeared in the April 1959 issue of *Amazing Stories* and reappears in various Smith collections, including *The Best of Cordwainer Smith* (New York: Nelson Doubleday, 1975).

p. 92: "with just a small laser dumped on a very small target."—Kammash, telephone interview, August 14, 2003.

p. 92: "in the inertial confinement fusion concepts we discussed in Chapter 3."—For more on the phenomenon of extreme light, see Gerard A.

Mourou and Donald Umstadter, "Extreme Light," *Scientific American*, May 2002, pp. 81–86.

p. 92: "in his proposal in a study for NASA's Institute for Advanced Concepts."—Terry Kammash, "Ultrafast-Laser Driven Plasma for Space Propulsion," NIAC 00-02 Final Report, October 2001. Available online at www.niac.usra.edu/studies. Kammash is also the editor of *Fusion Energy in Space Propulsion* (American Institute of Aeronautics and Astronautics, 1995).

p. 93: "then we'll have a system that could make it in a hundred years instead of a hundred thousand years."—Kammash interview, August 14, 2003.

CHAPTER 5

p. 96: "a space voyage of some forty years."—Smith's "The Lady Who Sailed the Soul" first appeared in the April 1960 issue of *Galaxy*.

p. 96: "finally fluttered out among the stars."—Smith's fiction has always been hard to gather in one place, but it is now available in *The Re-Discovery of Man: The Complete Short Fiction of Cordwainer Smith* (Framingham, Mass.: NESFA Press, 1993), 671 pages. The book's cover shows a three-sail interstellar spacecraft as envisioned by science fiction artist Jack Gaughan.

p. 97: "religious themes and allegories into much of his fiction."—Linebarger left an equally impressive list of nonfiction works, including the privately printed *Gospel of Sun Chung Shan* (Paris, 1932), *The Political Doctrines of Sun Yat Sen: An Exposition of the San Min Chu* (Baltimore: Johns Hopkins University Press, 1937), *The China of Chiang K'ai-shek: A Political Study* (Boston: World Peace Foundation, 1941), and *Psychological Warfare* (Washington, D.C.: Infantry Journal Press, 1948). Under the name Carmichael Smith, he wrote a highly regarded suspense novel called *Atomsk* (New York: Duell, Sloan & Pearce, 1949).

p. 97: "speeds high enough to reach the nearest stars."—A key study in this period was Forward's "Roundtrip Interstellar Travel Using Laser-Pushed Lightsails," *Journal of Spacecraft and Rockets* 21 (1984), pp. 187–95, but as we'll see, Forward's work was in constant transition as he injected new ideas into the mix.

p. 99: "confused its navigation sensors."—Donna Shirley, later manager of the Jet Propulsion Laboratory's Mars exploration program, saw Mariner 10 from the inside as a mission analyst. She tells its story in her book *Managing Martians* (New York: Broadway Books, 1998).

p. 99: "judicious tilting of the spacecraft solar panels."—"Mariner Venus/Mercury 1973 Status Bulletin #21," March 15, 1974, Mariner Venus/Mercury 1973 Project Office.

p. 100: "a long tradition that culminated in modern science fiction."—From an article in Salon, available online at http://www.salon.com/people/bc/2000/03/07/clarke.

p. 101: "dependent upon scientific theory and calculations."—Roger Bozzetto discusses the *Somnium* as the first known example of hard science fiction in "Kepler's Somnium, or, Science Fiction's Missing Link," in *Science Fiction Studies* #52, vol. 17, part 3 (November 1990).

p. 101: "an appendix stuffed with astronomical calculations."—The standard edition is Edward Rosen's *Kepler's Somnium* (Madison: University of Wisconsin Press, 1967).

p. 102: "you Galileo, for Jupiter."—As quoted in Wyn Wachhorst, *The Dream of Spaceflight* (New York: Basic Books, 2000), p. 6. Also see Arthur Koestler, *The Watershed: A Biography of Johannes Kepler* (Lanham, Md.: University Press of America, 1960), p. 195.

p. 102: "*Les mondes stellaires* (*Stellar Worlds*)."—Frederick Ordway III writes charmingly about *Aventures* and other early French science fiction works in his "Visions of the Moon: A Collector's Tale," which appeared in the November-December 2001 issue of *Ad Astra*.

p. 103: "Kepler's thesis about the displacement of comet's tails."—A good discussion of Lebedev's work is by Vassilis Lembessis, "P. N. Lebedev and Light Radiation Pressure," in *Europhysics News*, vol. 32, no. 1, 2001.

p. 104: "as it rushed past the sun."—Freeman Dyson, who draws frequently from Bernal in books like *Disturbing the Universe* (1979) and *Infinite in All Directions* (1988), wrote a wonderful essay exploring Bernal's contributions that was originally presented as a lecture in Bernal's honor at Birkbeck College, London. The printed version can now be found as an appendix in Carl Sagan, ed., *Communication with Extraterrestrial Intelligence* (Cambridge, Mass.: MIT Press, 1975).

p. 104: "designed to seed ideas among future authors."—Carl Wiley, "Clipper Ships of Space," *Astounding Science Fiction*, May 1951, pp. 135ff.

p. 105: "the beginning of modern work on the sail concept."—Richard Garwin, "Solar Sailing: A Practical Method of Propulsion within the Solar System," *Jet Propulsion* 28 (March 1958): 188–90.

p. 105: "one of the core articles in the field of interstellar studies."—"Solar Sail Starships: The Clipper Ships of the Galaxy" appeared in *Journal of the British Interplanetary Society* 34 (1981), pp. 371–80.

p. 105: "the force of the wind on the Earth's surface."—An excellent background piece on sail physics is available through one of JPL's Web sites at http://solarsails.jpl.nasa.gov/introduction/how-sails-work.html.

p. 107: "pushed by the pressure of light beams."—Pierre Boulle, *Planet of the Apes* (New York: Random House, 2000).

p. 108: "by the time you read this."—For more on the Ansari X Prize, see the official Web site at www.xprize.org.

p. 108: "the DLR German Aerospace Centre in Cologne."—"Solar Sails for Space Exploration—The Development and Demonstration of Critical Technologies in Partnership," *ESA Bulletin* 98 (June 1999).

p. 110: "the difficulties of deploying the latter."—Wright Friedman et al., "Solar Sailing: The Concept Made Realistic," American Institute of Aeronautics and Astronautics, Aerospace Sciences Meeting, Huntsville, Ala., Jan. 16–18, 1978. AIAA paper 78-82.

p. 110: "that would have been unrolled individually."—The heliogyro concept evolved from work by Richard MacNeal at the Astro Research Corporation and John Hedgepath in the mid-1960s.

p. 110: "not ready for that rendezvous,"—See Friedman's *Starsailing: Solar Sails and Interstellar Travel* for an overview of this project (New York: John Wiley & Sons, 1988).

p. 111: "a promising one."—Interview with Humphrey Price at the Jet Propulsion Laboratory, January 31, 2003.

p. 111: "the Solar Polar Imager."—Ibid.

p. 112: "we have to be circumspect."—Telephone interview with Moktar Salama, April 14, 2003.

p. 113: "missions outside the solar system."—Salama's recent laboratory work is discussed in his article "Ground Demonstration of a Spinning Solar Sail

Deployment Concept," *Journal of Spacecraft & Rockets* 40, no. 1 (January 2003): 9–14.

p. 114: "NASA flight validation mission."—See David M. Murphy and Paul A. Gierow, "Scalable Solar Sail Subsystem Design Considerations," presented at the 43rd Structures, Structural Dynamics and Materials Conference in Denver, Colo. in April 2002. Available online at http://www.aec-able.com/corpinfo/Resources/2002-1703-Murphy.pdf.

p. 115: "a 1945 article in the magazine *Wireless World*."—"Extra-Terrestrial Relays: Can Rocket Stations Give World-Wide Radio Coverage?" *Wireless World*, October 1945.

p. 115: "satellite television and communications platforms."—A true geostationary orbit puts the satellite directly over the equator, allowing it to remain stationary with respect to the ground observer, which is why these satellites are so useful for telecommunications and weather purposes.

p. 115: "while the Earth spins around underneath it."—Robert Forward, *Indistinguishable from Magic* (New York: Baen Books, 1995), 90.

p. 117: "impart thrust to a sail."—Benford's paper on the experiment is "Microwave Beam-Driven Propulsion Experiments for High-Speed Space Exploration," presented at the EuroEM 2000 conference in Edinburgh, Scotland. Myrabo's paper is "Experimental Investigation of Laser-Pushed Light Sails in a Vacuum," presented in 2000 at the Advanced Propulsion Conference at the Jet Propulsion Laboratory.

p. 117: "candidates for spaceflight."—"Sail Technology Beamed to Future Space Exploration," press release from the Jet Propulsion Laboratory's Media Relations Office, July 5, 2000.

p. 117: "I call my stories 'proposals.'"—Phone conversation with James Benford, February 19, 2003.

p. 118: "2,000 degrees Celsius in practical materials."—Ibid.

p. 118: "in a paper detailing the experiment."—James Benford, "Flight and Spin of Microwave-Driven Sails: First Experiments," *Proceedings Pulsed Power Plasma Science 2001*, IEEE 01CH37251, p. 548.

p. 119: "1,331 years."—Gregory Matloff, "The Perforated Solar Sail: Its Application to Interstellar Travel," *Journal of the British Interplanetary Society* 56 (2003): 255–61.

p. 119: "they'll take time to solve."—Interview with Les Johnson at Marshall Space Flight Center, July 30, 2003.

p. 120: "The flies can come through but you can't."—Benford phone conversation, February 19, 2003.

p. 120: "assisted by other mechanical means."—Gregory Benford et al., "Sail Deployment by Microwave Beam Experiments and Simulations," available online at http://www.physics.uci.edu/faculty/Sail_Dev_By_Mic_Beam.html.

p. 120: "a pack of cigarettes."—Greg Clark, "Breakthrough in Solar Sail Technology," space.com, March 2, 2000.

p. 121: "(some 30 million kilometers . . . well inside the orbit of Mercury)."—Charles E. Garner et al., "Lightweight Solar Sail for a Spacecraft Flying Near the Sun," NASA *Tech Brief* vol. 26, no. 10, from JPL Technology Report NPO-20854.

p. 121: "100 to 200 kilometers per second."—Price interview, January 31, 2003.

CHAPTER 6

p. 124: "while centrifugal force holds the spinning, web-slung mirror taut and flat in the void."—Eric K. Drexler, "The Canvas of the Night," in Arthur C. Clarke, ed. *Project Solar Sail* (New York: Roc, 1990), 44–45.

p. 124: "*The Starflight Handbook*."—Eugene Mallove and Gregory Matloff, *The Starflight Handbook* (New York: John Wiley & Sons, 1989). Their earlier article was "Solar Sail Starships—the Clipper Ships of the Galaxy," *Journal of the British Interplanetary Society* 34 (1981), pp. 371–80.

p. 124: "cables made of diamond to attach the payload to the sail."—For a more recent explanation of Matloff and Mallove's work, see Matloff's *Deep Space Probes* (Chichester, U.K.: Praxis Publishing Ltd, 2000).

p. 125: "metallic foils only a few hundred atoms thick have been created."—Eric K. Drexler, "High Performance Solar Sail Concept," *L5 News* 4 (May 1979): 7–9.

p. 125: "an ideal candidate for a mission this close to the Sun."—Mallove and Matloff, *Starflight Handbook*, 98.

p. 126: "and ride the laser beam all the way to the stars!"—Robert L. Forward, "Fast Forward Fifty Years," available on the Web at www.robertforward.com/Fast_Forward_Fifty_Years.htm.

p. 126: "*Science Digest* and *Galaxy Science Fiction*."—Robert L. Forward, "Pluto, Gateway to the Stars," *Missiles and Rockets* 10 (April 1962). Reprinted in *Science Digest*, August 1962, and later in *Galaxy Science Fiction*, December 1962. An indication of the continuing interest this article generated was yet another reprint, this time a revised version, in Jerry Pournelle, ed., *Black Holes* (New York: Fawcett Crest, 1978).

p. 126: "a novel called *Rocheworld* in which he laid out the essential elements that would guide his later research."—*Rocheworld* appeared in *Analog Science Fiction/Science Fact*, December 1982, and was later published in a longer version as *The Flight of the Dragonfly* (New York: Timescape, 1984). A final, still longer version appeared as *Rocheworld* in 1990 (New York: Baen Books).

p. 127: "a subsequent examination of the laser-driven sail concept by George Marx in *Nature* in 1966)."—G. Marx, "Interstellar Vehicle Propelled by Terrestrial Laser Beam," *Nature* 213 (July 1966): 22–23.

p. 127: "they ignored my advice and pretended it would work."—Forward, "Fast Forward Fifty Years."

p. 127: "but now it was a deceleration that would ultimately bring them to a stop at Barnard."—Forward, *Rocheworld*, 144–45.

p. 128: "P. C. Norem's 1969 paper analyzing ways of stopping a laser-pushed lightsail."—P. C. Norem, "Interstellar Travel: A Round Trip Propulsion System with Relativistic Velocity Capabilities," American Astronomical Society paper no. 69-388, June 1969.

p. 128: "whether the interstellar magnetic field was strong enough to provide the turning radius needed for the spacecraft."—As noted in Robert L. Forward, "Roundtrip Interstellar Travel Using Laser-Pushed Lightsails," *Journal of Spacecraft and Rockets*, 21 (March/April 1984): 187–95. Forward himself had studied curved interstellar trajectories in a 1964 paper "Zero Thrust Velocity Vector Control for Interstellar Probes: Lorentz Force Navigation and Circling," which ran in *American Institute of Aeronautics and Astronautics Journal* 2 (May 1964): 885–89.

p. 128: "in the technical publication *Journal of Spacecraft and Rockets* in 1984."—Forward, "Roundtrip."

p. 129: "is left for future generations to determine."—Ibid., 189.

p. 130: "and that means the beam has to be tightly focused."—Telephone interview with Edward Belbruno, April 18, 2003.

p. 132: "newer technologies that would overtake you wherever you are going."—Telephone interview with Freeman Dyson, May 29, 2003.

p. 132: "set the stage for Vostok and Apollo."—Greg Clark, "JPL Accomplishes Laser Sail First," space.com, March 1, 2000, at: http://www.space.com/businesstechnology/technology/laser_craft_001103.html

p. 133: "that speculated on the use of lasers to propel payloads into orbit."—Kantrowitz's papers are considered classics in the field, the major one being "Propulsion to Orbit with Ground-Based Lasers," in *Astronautics and Aeronautics* 10, no. 5 (May 1972), pp. 74–76. An earlier article in the same journal was "The Relevance of Space," *Astronautics and Aeronautics* 9, no. 3 (March 1971).

p. 133: "his 1985 book *The Future of Flight*, written with Dean Ing."—Leik Myrabo and Dean Ing, *The Future of Flight* (New York: Baen Books, 1985).

p. 133: "launch prices comparable to commercial airline tickets."—Telephone interview with Leik Myrabo, June 9, 2003.

p. 134: "mounted in a Boeing 747-400 aircraft and used to destroy incoming missiles during their boost phase."—"Northrop Grumman Delivers a High-Power Solid-State Laser for Missile Defense Agency's Airborne Laser Program," press release, Feb. 26, 2003.

p. 134: "according to Myrabo, who worked on a variant of the lightcraft idea for the group."—See Leik Myrabo et al., "Lightcraft Technology Demonstrator," Final Technical Report, prepared under Contract No. 2073803 for Lawrence Livermore National Laboratory and the SDIO Laser Propulsion Program, June 30,1989.

p. 134: "Building that space-based infrastructure is going to demand a quick way to space."—Myrabo interview, June 9, 2003.

p. 135: "by analyzing a wide range of possible substances, starting with metal sails."—An early paper on the subject is Landis's "Optics and Materials Considerations for a Laser-Propelled Lightsail," presented at the 40th International Astronautical Federation Congress, Malaga, Spain, Oct. 7–12, 1989 (IAA-89-664).

p. 135: "materials like diamond, silicon carbide, and zirconia."—Geoffrey A. Landis, "Small Laser-Propelled Interstellar Probe," presented at the 46th International Astronautical Federation Congress, Oslo, Norway, October 1995 (IAA-95-IAA.4.1.102).

p. 135: "it would heat to higher levels in the beam than he had originally calculated."—Geoffrey A. Landis "Dielectric Films for Solar and Laser-Pushed Lightsails," CP504, Space Technology and Applications International Forum-2000, edited by M. S. El-Genk (American Institute of Physics, 2000).

p. 135: "A bubble is transparent but you can see the colors."—Landis interview, April 3, 2003.

p. 135: "Robert Forward had proposed using diamond laser sails as early as 1986."—Robert L. Forward, "Laser Weapon Target Practice with Gee-Whiz Targets," *Proc. SDIO/DARPA Workshop on Laser Propulsion, July 7–18, 1986*, ed. Jordin Kare. Lawrence Livermore National Laboratory CONF-860778, vol. 2 (1987): 41–44.

p. 136: "as thin as 500 carbon atoms."—From "The World's Smoothest Diamond Films Pave the Path for Microscopic Motors," *Frontiers 2001*, Argonne National Laboratory. Available online at: http://www.anl.gov/OPA/frontiers2001/b4excell.html.

p. 136: "a future lightsail is thus well within the realm of the possible."—Paul W. May, "Diamond Thin Films: a 21st Century Material." *Philosophical Transactions of the Royal Society of London, Series* A (2000): 473–95.

p. 136: "it would use microwave power to return images from the encounter."—Robert L. Forward, "Starwisp: An Ultra-Light Interstellar Probe," *J. Spacecraft* 22 (1985b): 345–50.

p. 136: "the effect of an intense microwave beam on the materials from which the spacecraft was made."—Geoffrey Landis, "Advanced Solar- and Laser-Pushed Lightsail Concepts," Final Report for NASA Institute for Advanced Concepts, May 31, 1999. Available online at http://www.niac.usra.edu.

p. 136: "It would absorb all the power and vanish."—Landis interview, April 3, 2003.

p. 137: "the technology to demonstrate the use of microwaves for propulsion is available today."—Geoffrey A. Landis, "Microwave-Pushed Sails for Interstellar Travel," presentation to the 10th Advanced Propulsion Workshop, Huntsville, Alabama, April 5–8, 1999.

p. 137: "with interesting side effects that require further study."—In particular, "desorption," the out-gassing of materials in the sail as its temperature rises, can produce its own thrust, an effect not originally anticipated when Benford began his work.

p. 137: "beamed energy clearly works, i.e., needs no new physics, and has the most potential for near-term development."—Microwave Sciences, Inc., "Final Report: Laboratory Demonstration of Microwave Beamed Power Propulsion," Project Number R-700-200258-30025 (for Ohio Aerospace Institute).

p. 138: "the ramjet was not feasible."—As explained in Robert Zubrin's *Enter Space: Creating a Spacefaring Civilization* (New York: Tarcher/Putnam, 1999).

p. 138: "the magsail can be generated from within the spacecraft, eliminating spars or supporting materials."—D. G. Andrews and R. M. Zubrin, "Magnetic Sails and Interstellar Travel," International Astronautical Federation Paper IAF-88-5533, Bangalore, India, October 1988.

p. 139: "Winglee thinks velocities of 50 to 80 kilometers (31 to 50 miles) per second are possible."—Robert Winglee, "Mini-Magnetospheric Plasma Propulsion, M2P2," NIAC Award No. 07600-032. Final Report, November 2001.

p. 140: "We're figuring out how to do this ourselves."—Telephone interview with Robert Winglee, January 15, 2003.

p. 141: "What M2P2 can do in the solar system, we should be able to modify for interstellar work."—Landis interview, April 3, 2003.

p. 141: "nuclear bombs set off behind the vehicle to create the plasma that would drive a magnetic sail."—Dana Andrews and Robert Zubrin, "Nuclear Device-Pushed Magnetic Sails (MagOrion)," American Institute of Aeronautics and Astronautics Paper 97-3072, 33rd AIAA/ASME/SAE/ASEE Joint Propulsion Conference & Exhibit, Seattle, Washington, 1997.

p. 142: "before being positioned into the path of a particle beam."—Gregory Matloff, "Robosloth: A Slow Interstellar Thin-Film Robot," *Journal of the British Interplanetary Society* 49 (1996): 33–36. Also see his discussion of particle-beam propulsion in his book *Deep Space Probes* (Chichester, U.K.: Praxis Publishing, 2000), pp. 97–98.

p. 142: "vast electromagnetic launchers some tens of thousands of kilometers long."—Clifford Singer, "Interstellar Propulsion Using a Pellet Stream for Momentum Transfer," *Journal of the British Interplanetary Society* 33 (March 1980): 107–115.

p. 143: "such as lunar regolith or asteroids."—Nordley had written up these ideas in "Beamriders," a science article for *Analog Science Fiction and Fact*, vol. 119, no. 6 (July/August, 1999), and had discussed an earlier version in "Relativistic Particle Beams for Interstellar Propulsion," *Journal of the British Interplanetary Society* 46, no. 4 (1993), pp. 145–50. He added the self-steering pellet concept in a presentation at the 4th NASA advanced propulsion workshop at JPL in 1993 and extended it in a paper called "Interstellar Propulsion by Self-Steering Momentum Transfer Particles," presented at the 52nd International Astronautical Congress in Toulouse, France (IAA.4.1.05), October 2001.

p. 144: "an efficient design, more efficient than any lightsail."—Telephone interview with Gerald Nordley, May 12, 2003.

p. 145: "I started doing the calculations and realized that this made sense as a propulsion system."—Telephone interview with Jordin Kare, January 7, 2003.

p. 146: "the spacecraft is not coupled to the sails but driven by them."—Jordin T. Kare, "SailBeam: Space Propulsion by Macroscopic Sail-Type Projectiles," CP552, Space Technology and Applications International Forum-2001, edited by M. S. El-Genk (American Institute of Physics, 2001).

p. 146: "NASA's Institute for Advanced Concepts."—Jordin T. Kare, "High-Acceleration Micro-Scale Laser Sails for Interstellar Propulsion," Final Report, NIAC Research Grant #07600-070, revised February 15, 2002.

p. 146: "I love showing that slide."—Jordin Kare interview, January 7, 2003.

p. 147: "So it probably scales up better than most other schemes."—Ibid.

p. 149: "Galactic Matter and Interstellar Spaceflight."—Robert W. Bussard, "Galactic Matter and Interstellar Spaceflight." *Astronautica Acta* 6 (1960): 179–94.

p. 149: "given our current inability to drive a fusion reactor with them."—These numbers are drawn from R. H. Frisbee and S. D. Leifer, "Evolution of Propulsion Options for Interstellar Missions," a paper presented to the 34th Joint Propulsion Conference & Exhibit, July 13–18, Cleveland, Ohio. AIAA paper 98-3403, as cited in Landis, "The Ultimate Exploration."

p. 150: "some 4,000 kilometers in diameter, about the distance from Atlanta to San Francisco."—Carl Sagan, "Direct Contact among Galactic Civilizations by Relativistic Spaceflight," *Planetary and Space Science* 11 (1963): 485–98.

p. 150: "a novel by Poul Anderson called *Tau Zero* that made them ponder the seemingly impossible."—Poul Anderson, *Tau Zero* (New York: Doubleday, 1970).

p. 150: "If there was any single book that turned me on to actually engineering interstellar flight," Jordin Kare told me, "it was *Tau Zero*."—Jordin Kare interview, January 7, 2003

p. 151: "danced always on the same edge of disaster . . ."—Ibid., 43.

p. 152: "ramjet-propelled spacecraft became available to a much broader audience."—Carl Sagan and I. S. Shklovskii, *Intelligent Life in the Universe* (San Francisco: Holden-Day, 1966).

p. 152: "to accelerate the hydrogen into an exhaust stream."—Alan Bond, "An Analysis of the Potential Performance of the Ram Augmented Interstellar Rocket." *Journal of the British Interplanetary Society* 27 (1974): 674–685.

p. 152: "fuel provided by the ramjet's electromagnetic scoop."—See Gregory Matloff and Eugene Mallove, "Interstellar Flight: Aspects of Beamed Electric Propulsion," International Electric Propulsion Conference proceedings (A89-47426, 21-20), Garmisch-Partenkirchen, Federal Republic of Germany, Oct. 3–6, 1988 (Bonn: Deutsche Gesellschaft für Luft-, und Raumfahrt, 1988), 499–501.

p. 152: "you're trying to add energy to an already energetic mass stream."—Gerald Nordley interview, May 12, 2003.

CHAPTER 7

p. 157: "new dialogues between Achilles and the tortoise."—Carroll's work appeared as "What the Tortoise Said to Achilles," *Mind* 4, no. 14 (April 1895), pp. 278–80. The Hofstadter dialogue appears as introductory material to each chapter in the author's *Gödel, Escher, Bach: An Eternal Golden Braid* (New York: Basic Books, 1979).

p. 157: "then an antimatter rocket."—Interview with Marc Millis at Glenn Research Center, April 3, 2003.

p. 157: "to listen to their own words fifty years later reporting on their journey."—Van Vogt's tale is reprinted in the collection *Destination: Universe* (New York: Signet Books, 1952).

p. 158: "in his science fiction novel *Mars Crossing*."—Geoffrey Landis, *Mars Crossing* (New York: Tor Books, 2000).

p. 158: "And it's going to take a long time."—Interview with Geoffrey Landis at Glenn Research Center on April 3, 2003.

p. 159: "His 'Warp Drive When?' Web site."—http://www.grc.nasa.gov/WWW/PAO/warp.htm.

p. 160: "the proton is a whopping 430 times more massive than when it is at rest."—A wonderfully readable overview of the relationship between mass and energy is David Bodanis's *$E=mc^2$: A Biography of the World's Most Famous Equation* (New York: Walker & Co., 2000). Recall that mass and weight are not the same thing. Weight is measurable in a gravitational field, whereas in orbit, even the heaviest satellite has no weight. But all objects have mass, which is a measure of the resistance of the object to motion and is normally measured in kilograms. Mass is therefore a measure of the object's inertia. That orbiting satellite still has plenty of mass, so that moving it is no small job no matter what its weight. It was Einstein's insight that mass and energy are the same thing. Converting one to the other is possible and violates no physical laws. While each can change its state, the total amount of mass and energy remain the same. The conversion factor, of course, is huge. In Einstein's famous equation, the energy released is equal to the mass multiplied by the square of the speed of light.

p. 160: "7,000 times the mass of its payload in fuel."—Lawrence Krauss, as quoted in Jeff Greenwald, "To Infinity . . . and Beyond!" *Wired Magazine* issue 6.07, July 1998.

p. 161: "to achieve faster-than-light speeds."—NASA Office of Advanced Concepts and Technology, *Report from the Advanced Quantum/Relativity Theory Propulsion Workshop*, Jet Propulsion Laboratory, May 16–17, 1994.

p. 161: "*The Physics of Immortality*."—Tipler's book is breathtaking reading no matter what you think of his concept of using the energies of a so-called "big crunch" at the end of time to re-create everything that has gone before. The book was published by Doubleday in 1994.

p. 162: "I find words almost useless for mathematical thinking."—Roger Penrose, *The Emperor's New Mind: Concerning Computers, Minds, and the Laws of Physics* (New York: Oxford University Press, 1989), 548–49.

p. 163: "science essays and short stories."—Robert L. Forward, *Indistinguishable from Magic* (Riverdale, N.Y.: Baen Books, 1995).

p. 163: "time machines do not seem to be *forbidden* by the known laws of physics (this, however, does *not* mean they are allowed)."—Robert L. Forward, "Possible Faster-Than-Light Physical Phenomena: A Brief Survey," Advanced Quantum/Relativity Theory Propulsion Workshop, NASA Office of Advanced Concepts and Technology, May 16–17, 1994.

p. 164: "Ultrarelativistic Rockets and the Ultimate Fate of the Universe."—By A. Kheyfets and W. Miller, and F. Tipler III, respectively.

p. 164: "10 billion times the amount of energy in the entire universe."—Ulysses Torassa, "Warp Drives & Wormholes: It's a Workshop," *Cleveland Plain Dealer*, August 11, 1997.

p. 165: "far-out physics."—Lawrence Krauss, *The Physics of Star Trek* (New York: Basic Books, 1995). Krauss is also the author of *Beyond Star Trek: Physics from Alien Invasions to the End of Time* from the same publisher (1997).

p. 165: "they can still be useful by provoking other, more viable, ideas."—Marc G. Millis, "Breakthrough Propulsion Physics Workshop Preliminary Results," NASA TM-97-206241, Nov. 1997. Presented at Plenary Session III Views of Future. STAIF, Jan. 27, 1998, Albuquerque, N.M. Also Available in Space Technology and Applications International Forum-1998, ed. El-Genk, DOE CONF-980103, CP420, pp. 3–12.

p. 166: "a respected scientific journal like Classical and Quantum Gravity?"—*Quantum Gravity* 11 (May 1994): L73–L77.

p. 166: "the 'warp drive' of science fiction."—Ibid.

p. 167: "physicist John Cramer."—John G. Cramer, "The Micro-Warp Drive," *Analog Science Fiction and Fact*, February 2000.

p. 167: "the scientist got his inspiration from the show."—Alison Goddard, "Surfing to the Stars on Warped Space," *New Scientist* 142, issue 1929 (June 11, 1994): 18.

p. 167: "the possibility of engineering a similar effect by human efforts."—Alan H. Guth, *The Inflationary Universe* (New York: Perseus Publishing, 1997).

p. 167: "they sent forward to control the wall of the spacetime bubble in which they traveled."—Natario's paper, "Warp Drive with Zero Expansion," appears in *Classical and Quantum Gravity* 19, no. 6 (March 21, 2002), pp. 1157–65. See also Eugenie Samuel, "The Truth About Warp Drive," *New Scientist* 173, issue 2334 (March 16, 2002), p. 9.

p. 167: "seems to demand more energy than is available in the entire universe to make it work."—Michael J. Pfenning and L. H. Ford, "The Unphysical Nature of 'Warp Drive,'" *Classical and Quantum Gravity* 14 (1997): 1743–51.

p. 168: "How do you produce the exotic matter needed to manipulate negative energy?"—A straightforward look at the Alcubierre drive is provided by M.

Szpir in "Spacetime Hypersurfing," *American Scientist* 82 (Sept/Oct 1994), pp. 422–23. Also see Larry Gonick's column "Science Classics" for a playful look at the "warp and woof drive" in *Discover*, Dec. 1994, pp. 44–54.

p. 168: "a bubble of spacetime using warped space, one that is large on the inside but tiny on the outside."—Chris Van Den Broeck, "A 'Warp Drive' with More Reasonable Total Energy Requirements," *Classical and Quantum Gravity* 16 (1999), 3973–79.

p. 169: "This doesn't mean that the proposal is realistic."—Ibid., 3978.

p. 170: "We want people to think about the possibility of doing things in a different way."—Robert Cassanova, telephone interview, January 3, 2003.

p. 170: "Arthur C. Clarke's novel *The Fountains of Paradise?*"—Clarke's novel remains the definitive vision of a space elevator. It was first published by Harcourt in 1979.

p. 171: "the image of distant objects is magnified by intense gravity."—John Cramer et al., "Natural Wormholes as Gravitational Lenses," *Physical Review D* (March 15, 1995): pp. 3124–27.

p. 171: "a space completely filled with matter."—Bernard Haisch and Alfonso Rueda, "How to Abhor the Void While Loving the Quantum Vacuum," *Mercury* 29, issue 5 (Sept/Oct 2000): 32.

p. 172: "co-authored by physicist Alfonso Rueda."—Ibid.

p. 172: "the prestigious *Physical Review.*"—Bernard Haisch, Alfonso Rueda, and H. E. Puthoff, "Inertia as a Zero-Point Lorentz Force," *Physical Review* A, 49 (February 1994): 678–84.

p. 172: "sea of radiation that fills the entire universe."—Bernard Haisch, Alfonso Rueda, and H. E. Puthoff, "Beyond $E=mc^2$," *The Sciences*, November/December 1994, p. 27.

p. 173: "the interactions of charge and field."—Ibid., 26.

p. 174: "I don't know who the others were."—Arthur C. Clarke, *3001: The Final Odyssey* (New York: Ballantine Books, 1998), 61.

p. 174: "the even more fantastic possibility of controlling inertia."—Ibid., 245.

p. 174: "the Casimir Effect."—Robert Forward, "Extracting Electrical Energy from the Vacuum by Cohesion of Charged Foliated Conductors," *Physical Review B* 30, no. 4 (August 1984): 1770–73.

p. 175: "the real behavior of this quantum vacuum."—Jordan Maclay, "An Analysis of Vacuum Fluctuation Energy and Casimir Forces in Conductive Rectangular Cavities," *Physical Review* A 61, 052110-1 to 052110-18 (2000) provides technical background. "Energy Unlimited" by H. Bortman is a useful layman's description of Maclay's work in *New Scientist*, January 22, 2000, pp. 32–34.

p. 176: "there would be no such thing as inertia."—Telephone interview with James Woodward, October 7, 2003.

p. 176: "transient fluctuations in its mass."—See James F. Woodward and Thomas Mahood, "Mach's Principle, Mass Fluctuations, and Rapid Spacetime Transport." Available online at http://chaos.fullerton.edu/~jimw/staif2000.pdf.

p. 179: "it would be our first evidence that wormholes actually exist."—The paper Landis is referring to is Cramer et al., "Natural Wormholes."

p. 180: "stretching to infinity in both directions."—Geoffrey L. Landis, "Approaching Perimelasma," in *Impact Parameter* (Urbana, Ill.: Golden Gryphon Press, 2001).

p. 180: "cards home to be shuffled."—"What We Really Do Here at NASA" also appears in *Parameter.*

CHAPTER 8

p. 184: "for Internet communications."—Ben Iannotta, "Earth, You've Got Mail," *New Scientist* 162, issue 2187 (May 22, 1999): 32.

p. 184: "one-tenth the speed of a broadband Internet connection."—David F. Carr, "The 100 Million Mile Network," *Baseline* no. 27 (February 2004), 24–26.

p. 184: "a tenth of that is likely to be usable."—Telephone interview with James Lesh, December 19, 2002.

p. 186: "Voyager could deliver from Neptune."—Lesh cites this figure in Marvin J. Wolf, "From Rocketeers to Solar Sailors: At the Jet Propulsion Laboratory, Scientists Are Building Robots to Explore the Solar System—and Even Making Plans to Sail to the Stars," *Los Angeles Times Magazine*, November 14, 1999.

p. 187: "to Earth by radio."—A. T. Lawton and P. P. Wright, "Project Daedalus: The Vehicle Communications System," *Project Daedalus Final Report, Journal of the British Interplanetary Society* (1978), pp. S163–S171.

p. 187: "deep-space missions in coming decades."—Keith A. Beals et al., "Project Longshot: An Unmanned Probe to Alpha Centauri," NASA/USRA University Advanced Design Program Project Report for 1987–1988, U.S. Naval Academy.

p. 190: "into a much tighter beam."—Robert Zubrin offers a clear explanation of this and many other issues in *Entering Space* (New York: Tarcher/Putnam, 1999), pp. 260–61.

p. 190: "by a factor of four or five."—T. B. H. Kuiper and G. M. Resch, "Deep Space Telecommunications," from *Perspectives on Radio Astronomy: Technologies for Large Antenna Arrays*, proceedings of a conference held at the ASTRON Institute in Dwingeloo, Netherlands, April 12–14, 1999. Available online at http://www.astron.nl/documents/conf/technology/tech08w.pdf. See also "Space Probes Could Talk Faster with Light," UPI Science News, July 25, 2002, for the upgrade to Ka-band.

p. 190: "a change of paradigm may be necessary."—Alex Harwit, Martin Harwit, and Joss Bland-Hawthorn, "Laser Telemetry from Space," *Science* 297, no. 5581 (July 26, 2002): 523.

p. 191: "communications during peak periods."—H. Hemmati, "Free-Space Optical Communications at JPL/NASA," available online at http://lasers.jpl.nasa.gov/PAPERS/REVIEW/overview.pdf.

p. 191: "*Journal of the British Interplanetary Society*."—James Lesh, C. J. Ruggier, and R. J. Cesarone, "Space Communications Technologies for Interstellar Missions," *Journal of the British Interplanetary Society* 49 (1996): 7–14.

p. 191: "the communications system is not a problem."—Lesh, telephone interview, December 19, 2002.

p. 191: "volume aboard the spacecraft."—Ben Iannotta, "That's Entertainment," *New Scientist* 169, issue 2283 (March 24, 2001): 32.

p. 193: "a laser beacon sent from Earth."—For more on the Optical Communications Demonstrator, see Multhu Jeganathan and Jim Lesh, "Laser Communications Unit Prepares for Test Flight," *Laser Focus World* 35, no. 6 (June 1999), pp. 103–109.

p. 194: "a distance of six million kilometers (3,725,000 miles)."—Mark Whalen, "GOPEX Reaches Galileo via Laser Beam," *JPL Universe*, December 30, 1992.

p. 194: "Earth's atmosphere upon the signal."—K. Wilson, M. Jeganathan, and J. R. Lesh, "Results from Phase-1 and Phase-2 GOLD Experiments," TDA *Progress Report* 42-128, February 15, 1997.

p. 194: "is caused by the Earth's atmosphere."—Bruno von Wayenburg, "Europe to Test Laser Link to Moon Probe," *New Scientist* 178, issue 2395 (May 17, 2003).

p. 194: "95 percent of the time."—See the Jet Propulsion Laboratory's background information on laser communications, available online at lasers.jpl.nasa.gov.

p. 194: "former director, Edward C. Stone."—Edward C. Stone, "Communications Technologies for Space Exploration," *Proceedings of the IEEE* vol. 87, no. 6 (June 1999). See also Stelzried Edwards et al., "NASA's Deep Space Telecommunications Roadmap," from JPL's Telecommunications and Mission Operations Directorate, available online at http://www.nasda.go.jp/pr/event/app/spaceops/paper98/track1/1b029.pdf.

p. 194: "optical maser," or laser."—Charles Townes and Arthur Schawlow, "Infrared and Optical Masers," *Physical Review* no. 112 (December 1958): 1940–49.

p. 195: "the search for extraterrestrial signals."—Giuseppe Cocconi and Philip Morrison, "Searching for Interstellar Communications," *Nature* 184, no. 4690 (September 19, 1959): 844–46.

p. 195: "with R. N. Schwartz in *Nature*."—R. N. Schwartz and C. H. Townes, "Interstellar and Interplanetary Communication by Optical Masers," *Nature*, 190, no. 4772 (April 15, 1961): 205–208.

p. 200: "a common expensive resource for long-haul links."—Telephone interview with Adrian Hooke, February 27, 2003.

p. 200: "not by radio but by gasoline."—Ibid.

p. 202: "a similar system called Galileo."—A description of the European GPS system appears in "Europe's Answer to GPS Irks Pentagon," *New Scientist* 178, issue 2398 (June 7, 2003).

p. 203: "the delicate navigation needed on the approach."—NASA press release, "Stardust Status Report," May 11, 2001.

p. 203: "the motion of local stars and gas."—John H. Mauldin reviews these numbers in his *Prospects for Interstellar Travel*, published by the American Astronautical Society (San Diego: Univelt, 1992), p. 167.

p. 204: "from 520 to 603 light years."—Science Applications International Corporation, "Interstellar Spaceflight Primer," NASA Contract No. NASW-5067, February 2001, p. 71.

p. 206: "pushing up close to light speed."—Saul Moskowitz, "Trans-Stellar Space Navigation," *American Institute of Aeronautics and Astronautics Journal* 6 (1968): 1021–29; R. W. Stimets and E. Sheldon, "Celestial View from a Relativistic Starship," *Journal of the British Interplanetary Society* 34 (1981): 83–99; E. Sheldon and R. H. Giles, "Celestial Views from Non-Relativistic and Relativistic Interstellar Spacecraft," *Journal of the British Interplanetary Society* 36 (1983): 99–114.

p. 207: "a sight no one may ever see."—Mauldin, *Prospects*, 181.

p. 207: "the spacecraft's direction of travel."—Ibid., 170.

p. 208: "in the infrared and ultraviolet regions."—Michael Dornheim, "Deep Space 1 Prepares to Launch Ion Drive," *Aviation Week & Space Technology* (October 5, 1998): 108.

p. 208: "more important science data, including photos."—"NASA's Deep Space One Succeeds in Close Asteroid Flyby," from a NASA/JPL press release, July 29, 1999.

p. 209: "DS1's mission data system in 1998."—LaForge worked on Deep Space 1's flight software as a NASA Faculty Fellow at JPL in 1997. His observations appear in a letter to MIT's *Technology Review*, published as "Mis-Impressions about NASA," which was posted to the *Review's* Web site on March 7, 2003, and were later conveyed in a telephone interview with the author on July 22, 2003.

p. 209: "we won't know how well things work."—Laurence LaForge in an e-mail to the author, July 22, 2003.

p. 209: "as spacecraft learn how to set their own courses."—ESA Press Release, "Set Your Own Course for the Stars," November 12, 2002.

p. 210: "based on the center of the galaxy rather than the galaxy as seen from Earth."—These musings draw largely on astronomer Sten Odenwald's speculations, found online at http://www.astronomycafe.net/qadir/q599.html.

CHAPTER 9

p. 212: "one of its ropy tentacles."—"Taklamakan," which won Sterling a Hugo Award in 1999, was originally published in *Asimov's Science Fiction* (October/November 1998).

p. 212: "the extinction of the human race."—For more on Hans Moravec's views of robotics, see his book *Robot: Mere Machine to Transcendent Mind* (New York: Oxford University Press, 1998).

p. 212: "gray goo."—Joy's warnings appear in the April 2000 issue of *Wired* magazine in the article "Why the Future Doesn't Need Us." The Drexler reference is to his book *The Engines of Creation* (New York: Doubleday, 1986). The section called "Dangers and Hopes" is particularly insightful.

p. 215: "a high degree of mobility and autonomy."—T. J. Grant, "Project Daedalus: The Need for On-Board Repair," in A. R. Martin, ed., *Project Daedalus Final Report*. Supplement to the *Journal of the British Interplanetary Society*, 1978, pp. S172–S179.

p. 215: "the ship's fusion drive."—As noted in Alan Bond and Anthony R. Martin, "Project Daedalus: The Mission Profile," in A. R. Martin, ed. *Project Daedalus Final Report*, p. S41.

p. 216: "to choose an alternative method of repairing the crack, or some acceptable combination."—T. J. Grant, "Project Daedalus: The Computers," *Project Daedalus Final Report*, p. S141.

p. 216: "monitoring onboard oxygen and pressure, perhaps, or making minor repairs."—Catherine Zandonella, "It's Tiny, It's Round, and Every Astronaut Should Have One," *New Scientist* 163, issue 2195 (July 17, 1999), p. 7. See also "Smart Robot Orbs to Aid Space Crews," *Science News* 156, issue 13 (September 25, 1999), p. 197.

p. 216: "the Daedalus warden in almost human shape."—Duncan Graham-Rowe, "Robonaut Takes the Risks So You Don't Have To," *New Scientist* 163, issue 2205 (September 25, 1999), p. 20.

p. 218: "result in increasing sophistication of the resulting self-designed systems."—Telephone interview with Jordan B. Pollack, November 5, 2003.

p. 219: "locomoting robots called genobots."—For more on generative encodings and genobots, see Jordan B. Pollack et al., "Three Generations of Automatically Designed Robots," *Artificial Life* 7, no. 3 (2001): 215–23.; also see Jordan B. Pollack et al., "Computer Creativity in the Automatic Design of Robots," *Leonardo, Journal for the International Society for Arts Sciences and Technology* 36, no. 2 (2003): 115–21. The DEMO Web site at Brandeis offers background documents and updates to current work at www.demo.cs.brandeis.edu.

p. 219: "we virtually build the blueprints."—Jordan B. Pollack, telephone interview, November 5, 2003.

p. 220: "galactic distances ultimately also must be developed."—R. A. Freitas and W. P. Gilbreath, "Advanced Automation for Space Missions," NASA Report CP-2255, 1982.

p. 220: "the next series of operations."—For more details on this aspect of Deep Space 1's mission, see John McHale, "Remote Agent Software on NASA's Deep Space 1 Spacecraft Is a Success," *Military & Aerospace Electronics*, 10, issue 9 (September 1999), p. 1.

p. 221: "lacks the planning capabilities of Deep Space 1's Remote Agent."—These details come from Douglas E. Bernard and Barney Pell, "Designed for Autonomy: Remote Agent for the New Millennium Program," Proceedings of the Fourth International Symposium on Artificial Intelligence, Robotics, and Automation for Space, 1997.

p. 222: "controlling a robot that used evolving circuits to avoid walls."—Adrian Thompson, *Hardware Evolution: Automatic Design of Electronic Circuits in Reconfigurable Hardware by Artificial Evolution*, Distinguished Dissertation Series (London: Springer-Verlag, 1998).

p. 223: "HAL-9000 of Arthur C. Clarke's design, immortalized in Stanley Kubrick's film *2001*, was the product of the same University of Illinois at Urbana-Champaign."—Clarke made HAL's point of origin clear in a 1997 letter to UIUC, saying he chose the school because his old math professor, George McVittie of Kings College, London, spent his last years there.

p. 224: "an operating system called Viva to continually reconfigure itself to deal with new situations."—Brice Wallace, "Hyper-Drive for Computers," *Deseret Morning News*, June 9, 2003.

p. 224: "for measuring the software's performance."—In a letter to the MIT *Technology Review* in the spring of 2003, LaForge wrote ". . . we would have a much better sense of how good (or bad) our autonomic computing is, were we to properly instrument our software. Ironically, there is plenty of research in place to do exactly this, with much of it due to NASA's own experts . . . But NASA has never had the discipline or common sense to properly apply this research . . . Until NASA makes it a priority to properly instrument software . . . claims of improvements in reliability fall to the dustbin of snake oil remedies."

p. 225: "as long as 600 years."—LaForge's NIAC study is available at the agency's Web site: www.niac.usra.edu. Its title is "Architectures and Algorithms for Self-Healing Autonomous Spacecraft," a Phase I study that ran from May 1, 1999, to October 31 of the same year.

p. 225: "the star probe's projected 10,000 bps."—Ibid.

p. 226: "eventually make it incapable of communicating."—For the immune system comparison, see A. Avizienis, "The Hundred Year Spacecraft," Proceedings of the First NASA/DOD Workshop on Evolvable Hardware, Jet Propulsion Laboratory, Pasadena, Calif., July 19–21, 1999, pp. 233–39.

p. 226: "the bit-level analogy of a common cold."—LaForge, "Architectures and Algorithms," 14–15.

p. 227: "unrelated failure in a power system control module."—Press release, "Diagnostic Software to Keep Launch Vehicles Healthy," Marshall Space Flight Center, July 8, 2002.

p. 227: "with nanomachines fabricating repair materials out of nearby molecules in the hull."—Kevin Bonsor, "How Self-Healing Spacecraft Will Work," posted on HowStuffWorks at www.howstuffworks.com.

p. 227: "about 7 percent of the mass of the payload, unacceptably high for interstellar work."—See John H. Mauldin's *Prospects for Interstellar Travel*, vol. 80 of the American Astronautical Society's Science and Technology Series (San Diego: Univelt, 1992), p. 193. Mauldin cites T. J. Grant's "Need for Onboard Repair," in the *Journal of the British Astronomical Society's* Project Daedalus supplement, 1978.

p. 228: "it has an Earth-like atmosphere, continents, and oceans of liquid water."—Bear's *Queen of Angels* was published by Warner Books in 1990.

p. 229: "It will have to make judgments heretofore reserved for human beings."—Ibid., 81.

p. 229: "he means the creation of superhuman intelligence that may end the human era."—Vinge's original paper on the Singularity is widely available on the Internet at sites like http://www.ugcs.caltech.edu/~phoenix/vinge/vinge-sing.html.

p. 230: "in the form of future-looking reproductive efficiency."—Jordan B. Pollack interview, November 5, 2003.

p. 230: "to survive and perform at optimal functionality during long duration in unknown, harsh or changing environments . . ."—Adrian Stoica, Didier Keymeulen, and Jason Lohn, preface to proceedings of the First NASA/DOD Workshop on Evolvable Hardware, July 19–21, 1999.

p. 230: "Adrian Thompson's FPGAs actually worked in an analog fashion."—For more on this, see Clive Davidson's "Creatures from Primordial Silicon," *New Scientist* 156, issue 2108 (November 15, 1997), p. 30. Writing in 1998, Thompson described the analog behavior this way (the XC6216 is the FPGA chip from Xilinx used in the experiment): "It is operating as a continuous-time analogue dynamical system, even though the XC6216 is intended to be used as a digital device. The design constraints of a digital methodology were not imposed, so evolution has used whatever behaviours the chip manifested." This is from Thompson's "Exploring Beyond the Scope of Human Design: Automatic Generation of FPGA Configurations through Artificial Evolution," extended abstract, 8th Annual Advanced PLD & FPGA Conference, 1998.

p. 232: "you are moving toward genuine autonomy."—The work of Stoica, Keymeulen, and Zebulum is deftly explained in Anil Ananthaswamy's article "Space Babies," which ran in *New Scientist* 169, issue 2276 (February 3, 2001), p. 26.

p. 232: "This has the potential to largely enhance the capabilities of future space systems."—Adrian Stoica, "Evolvable Hardware: From On-Chip Circuit Synthesis to Evolvable Space Systems," in Proceedings of the 30th IEEE International Symposium on Multiple-Valued Logic, May 23–25, 2000.

p. 232: "You go simply because you are expendable."—Cordwainer Smith, "The Lady Who Sailed the Soul," in *The Best of Cordwainer Smith* (New York: Nelson Doubleday, 1975), p. 57.

CHAPTER 10

p. 236: "nuclear pulse engines and solar sails."—A. R. Martin, "World Ships: Concept, Cause, Cost, Construction and Colonization," *Journal of the British Interplanetary Society* 37 (1984): 243–53.

p. 237: "There is a way out if you are smart."—Telephone interview with Gregory Matloff, August 18, 2003.

p. 238: "if, within the next sixty years, the gross world product rose to a per capita level like that of the United States today, then such a huge figure might be justifiable."—Curt Mileikowsky, "How and When Could We Be Ready to Send a 1,000 KG Research Probe with a Coasting Speed of 0.3c to a Star?" *Journal of the British Interplanetary Society* 49 (1996): 335–44. The proceedings of the "Interstellar Robotic Probes: Are We Ready?" conference were published in their entirety in JBIS.

p. 238: "probes to all stars within twenty light years could be launched during the first few months of operations."—Dana G. Andrews, "Cost Considerations for Interstellar Missions," *Journal of the British Interplanetary Society* 49 (1996): 123–28.

p. 239: "An interstellar probe is feasible with technology known today."—Landis's paper is emerging as a classic in the field. Its title is "Small Laser-Propelled Interstellar Probe," presented at the 46th International Astronautical Congress in Oslo, Norway, October 1995, paper IAA-95-IAA.4.1.102. It should be noted that Landis has subsequently changed his thinking on the beryllium sail he suggests in this paper. He is now investigating dielectric films including diamond.

p. 240: "will no doubt misjudge where the most successful applications are."—Ed Gerstner, "World's Smallest Electric Rotor Made," *Nature Science Update*, July 24, 2003. For the full article, see A. M. Fennimore et al., "Rotational Actuators Based on Carbon Nanotubes," *Nature* 424 (2003): 408–410.

p. 241: "as agile as a hummingbird with a brain weighing no more than a gram."—Dyson develops this concept in his book *Infinite in All Directions* (New York: Harper & Row, 1988), p. 197. This book is based upon a series of lectures Dyson gave in Aberdeen, Scotland, in 1985.

p. 241: "crossing in cryogenic storage."—See Matloff's *Deep Space Probes* (New York: Springer-Verlag, 2000), p. 62.

p. 241: "an analysis of how machines may one day physically reproduce."—*Kinematic Self-Replicating Machines* will be published in 2004 by Landes Bioscience, Georgetown, Texas. Volumes I and IIA of *Nanomedicine*, also published by Landes Bioscience, are currently available and supported by a Web site: www.nanomedicine.com.

p. 241: "his study of medical nanorobots was the first ever to be published in a peer-reviewed mainstream biomedical journal."—Robert A. Freitas Jr., "Exploratory Design in Medical Nanotechnology: A Mechanical Artificial Red Cell," *Artificial Cells, Blood Substitutes, and Immobil. Biotech.* 26 (1998): 411–30.

p. 242: "you could practically push them with a flashlight."—Interview with Ed Belbruno, April 18, 2003.

p. 243: "you could put the equivalent of thousands of human brains worth of intelligence even with a massive backup of redundant systems."—Interview with Robert Freitas, July 24, 2003.

p. 243: "a self-reproducing probe called REPRO."—An earlier discussion of self-reproducing probes can be found in Nigel Calder's *Spaceships of the Mind* (New York: Viking Press, 1978), p. 89, which draws on earlier comments by Freeman Dyson. Calder considers the technology in relation to development of the solar system. Chris Boyce also considers such probes in his *Extraterrestrial Encounter: A Personal Perspective* (Secaucus, N.J.: Chartwell Books, 1979), but as far as I know, Freitas was the first to subject self-reproducing probes to rigorous scientific analysis.

p. 244: "A few such probes sent from our solar system could spread throughout the entire Milky Way without further cost or intervention."—Robert A. Freitas Jr., "A Self-Reproducing Interstellar Probe," *Journal of the British Interplanetary Society* 33 (1980): 251–64. Also available in revised form on the Web at http://www.rfreitas.com/Astro/ReproJBISJuly1980.htm.

p. 244: "a self-replicating lunar factory system that could one day develop into a series of automated interstellar probes."—The proceedings for the 1980 NASA study, which was co-sponsored by the American Society for Engineering Education and held in Santa Clara, California, can be found as "Advanced Automation for Space Missions," NASA Conference Publication 2255 (NASA Scientific and Technical Information Branch, 1982). Also available in partial form online at http://www.islandone.org/MMSG/aasm.

p. 245: "replication times on the order of weeks."—Interview with Robert Freitas, July 24, 2003.

p. 245: "self-reproduction to maximize the information flow."—F. Valdes and Robert A. Freitas Jr., "Comparison of Reproducing and Nonreproducing Starprobe Strategies for Galactic Exploration," *Journal of the British Astronomical Society* 33 (1980): 402–406.

p. 245: "Hungarian-born mathematician and computer savant John von Neumann."—The classic paper in question is von Neumann's *Theory of Self-Reproducing Automata*, edited and completed by A. W. Burks (Urbana, Ill.: University of Illinois Press, 1966). Michael Arbib examined these views three years later in his *Theories of Abstract Automata*, Prentice-Hall Series in Automatic Computation (Englewood Cliffs, N.J.: Prentice Hall). A later distillation and extension of these thoughts came with Chris Boyce's *Extraterrestrial Encounter: A Personal Perspective* (Secaucus, N.J.: Chartwell Books, 1979).

p. 245: "*Galaxy Science Fiction* magazine."—Reprinted in *The Collected Stories of Philip K. Dick, Volume 4* (New York: HarperCollins, 1987).

p. 246: "Where are they?"—See Tipler's "Extraterrestrial Intelligent Beings Do Not Exist," *Quarterly Journal of the Royal Astronomical Society* 21 (1980), pp. 267–81. And for a spirited rebuttal, see Carl Sagan and William Newman's "The Solipsist Approach to Extraterrestrial Intelligence," *Quarterly Journal of the Royal Astronomical Society* 24 (1983), p. 113. Sagan and Newman argue that the spread of self-reproducing probes would be so viral-like that they would endanger the very existence of the Galaxy and would therefore never be built.

p. 247: "They are among us and they call themselves Hungarians."—Stephen Webb discusses numerous possible answers to the Fermi Paradox in his lively book *Where Is Everybody?* (New York: Copernicus Books, 2002).

p. 247: "probes would be ideal for making contact with other intelligent races."—R. N. Bracewell, "Communications from Superior Galactic Communities," *Nature* 186 (1960): 670–71. Reprinted in A. G. Cameron, ed., *Interstellar Communication* (New York: W. A. Benjamin, 1963), 243–48.

p. 247: "the subject of radio SETI had appeared the previous year in *Nature*."—Philip Morrison and Giuseppe Cocconi, "Searching for Interstellar Communications," *Nature* 184, no. 4690 (September 19, 1959): 844–46.

p. 248: "searches for interstellar probes in our own solar system have been carried out at Kitt Peak National Observatory and the University of California at Berkeley, though without result."—One fascinating proposal, advanced by Scottish writer Duncan Lunan in 1974, was that a series of unusual radio receptions in the 1920s that seemed to be echoes of earlier transmissions might have been produced by a probe in our solar system. Later arguments against the idea seem persuasive.

p. 248: "halo orbits" around the Lagrangian points."—The halo orbits are Robert Freitas's idea. Freitas conducted a series of observations of these areas in 1979 using a 30-inch telescope at Leuschner Observatory in Lafayette, California. His work was reported in "A Search for Natural or Artificial Objects Located at the Earth-Moon Libration Points," *Icarus* 42 (1980), pp. 442–47. Freitas argues in the paper that the Earth/Moon Lagrangian points are not in fact stable, being disturbed by solar gravity, but that large, stable orbits around these points do exist.

p. 248: "the emergence of a technological civilization."—A useful overview, and a set of principles regarding what humanity might do if contact with such a probe were achieved, is found in Allen Tough's "Small Smart Interstellar Probes," *Journal of the British Astronomical Society* 51, no. 5 (May 1998), p. 167 ff.

p. 248: "ethical issues, many of which Robert Freitas has analyzed."—See, for example, Freitas's "The Legal Rights of Extraterrestrials," *Analog Science Fiction and Fact* (April 1977), pp. 54–67. He also discusses the issues in "Metalaw and Interstellar Relations," *Mercury* 6 (March/April 1977), pp. 15–17.

p. 249: "short story collections and novels, a series that continues today."—The first Berserker story I know about is "Fortress Ship," which appeared in *IF's* January 1963 issue. It became part of the collection *Berserker* (New York: Ballantine Books, 1967), which collected eight short stories from this period.

p. 249: "into the galactic community."—Ferris discusses the ramifications of all this, and the intriguing similarities between such a galaxy-spanning network and the human brain itself, in *The Mind's Sky: Human Intelligence in a Cosmic Context* (New York: Bantam Doubleday Dell, 1992).

p. 250: "Campbell could argue his readers would become suspicious if such stories were *not* to appear."—This oft-repeated story is told in Albert Berger's *The Magic That Works: John W. Campbell and the American Response to Technology* (San Bernardino, Calif.: Borgo Press, 1993).

p. 252: "Nobody is going to put his job on the line by putting in new systems that may fail."—Steve Lohr, "Computers Driving Shuttle Are to Be Included in Inquiry," *New York Times*, February 7, 2003, p. A22.

p. 253: "men whose visions no governmental project could encompass."—Freeman Dyson, *Disturbing the Universe* (New York: Harper & Row, 1979), 117.

p. 253: "but sometimes the paperwork is overwhelming."—As quoted in Ben R. Finney and Eric M. Jones, *Interstellar Migration and the Human Experience* (Berkeley: University of California Press, 1985), 13.

p. 253: "And a people descended from supreme Jupiter."—*Aeneid* 1.378–380, as translated by my friend Barney Rickenbacker.

p. 253: "We often search for no practical reason."—Charles Pasternak, *Quest: The Essence of Humanity* (New York: John Wiley & Sons, 2003).

p. 255: "using a voyaging technology they alone possessed to sail where no one had ever been before."—Ben R. Finney, "Ocean Space," in *Interstellar Migration*, p. 169.

INDEX